Wildlife and Disease in India

Wildlife and Disease in India

Arnold C. Long

CYBER TECH PUBLICATIONS
4264/3, Ansari Road, Daryaganj, New Delhi-110002 (India)
Ph.: 011-23244078. Fax: 011-23280028
E-Mail: cyberpublicationsdelhi@yahoo.com

WILDLIFE AND DISEASE IN INDIA

Arnold C. Long

First Edition 2009

Published by :

G.S. Rawat for **Cyber Tech Publications**
4264/3, Ansari Road, Daryaganj, New Delhi-110002 (India)
Ph.: 011-23244078. Fax: 011-23280028
E-Mail: cyberpublicationsdelhi@yahoo.com

Printed at : **Amit Enterprises**, Maujpur, Delhi, (9810878184)

Preface

This is a brief introductory book on the wildlife disease in India. A large number of people and activists has been, over the years, positively involved in conservation of wildlife and wetlands of the southern districts of Tamil Nadu through ecological research and environment education activities. Koonthakulam Bird Sanctuary is one of the study areas of IRNE where several migratory and non-migratory water birds co-exist from October to April every year. Since there is considerable speculation that some of the migratory birds coming from North-eastern territories could bring Bird Flu disease into India, it was considered necessary and timely to use expertise to conduct this awareness and training Workshop exclusively for officials of some of the government departments who are directly responsible for handling the Bird Flu situation. This effort is meant to complement the good work being already done by the above departments. The active cooperation extended to IRNE by the Tamil Nadu Forest Department and the Animal Husbandry Department contributed to the success of this programme.

This book is about the problems and diseases in wildlife in India.

—*Author*

CONTENTS

Preface

1. Introduction 1
2. Wildlife Protection Act 19
3. Prohibition of Animal Trade 58
4. Wildlife Specified Plants 83
5. Import-export Policy 99
6. The Indian Forest Act 121
7. Name of the State/Union Territory Act 155
8. Ministry of Environment and Forests 165
9. Forest Conservation Rules 178
10. The Wild Life Protection Amendment 222

Bibliography 260

Index 262

1

INTRODUCTION

Animals as well as plants play a significant role in human life. They have vital medicinal, ecological, agricultural, commercial and recreational importance. For instance, those suffering from heart ailments will die within 72 hours of a heart attack if not provided with digitalis, a drug obtained from the purple bluefox. Also the right balance between humans, plants and animals lead to a healthy ecosystem. Wildlife in India significantly contributes to the growth of regional economies. Hence, the importance of conserving endangered species is immense.

CAUSES OF ENDANGERMENT

The primary cause of endangerment is the destruction of natural habitat. With continuing ecological changes, habitats also undergo regular alterations. The release of greenhouse gases brings about changes in the global climate and has resulted in the loss of microbes in the soil which in turn has made the tropical forests subject to destruction and the extinction of many wildlife and aquatic animals. A survey says that India has lost 26,000 square kilometers of dense forest and 7,000 square kilometers of scrub due to shifting cultivation, encroachment and felling.

Human activity has heavily affected the individual species and habitats. The unwanted interaction between native animals and interlopers, primarily initiated by humans, result in harmful consequences. Overexploitation of animals by humans is another important cause of endangerment. Unlimited whaling resulted in the extinction of a number of species of whales. However, the universal cause of endangerment will remain till humans make an effort to protect the natural environments where the animals dwell. India has several wildlife sanctuaries and national parks most of which were established in an attempt to conserve the endangered animals. In the wildlife reserves, the threatened animals of India are protected from animals, natural hazards and other causes of endangerment.

The stories of William Stockley, Kinloch and other travelers of a century or more ago painted a romantic view of the hunting adventures and wildlife species encountered in the highlands of central Asia and Trans-Himalaya. The world has changed dramatically since those days and my own experience has been a curious mixture that could be termed a "post-modern" traveler's glimpse of and the initiation of a conservation movement in the trans-Himalaya – actually, just some occasional views of these changes through a foreigner's eyes. Following the exploits of the British and other western hunters in the last century various military actions coupled with general increases in road access and firearm technology have rendered the widespread wildlife but a remnant of its former abundance. Species of the most inaccessible areas, such as blue sheep and in places ibex, probably did not have significant population reductions in those sites. But others such as the Ladakh urial, Tibetan gazelle, Tibetan argali were heavily hunted.

Today, with hunting restrictions, we are apparently seeing a recovery of some of these wild ungulate populations, for example the Ladakh urial along the Indus and both blue sheep and ibex in accessible parts of their range, as well as the return of a few gazelle in eastern Ladakh. But in general for ungulates of the open highlands, a modernising animal husbandry leaves little room for species of the plains and open hills (such as gazelle, argali, wild yak). The conservation of these will require significant commitment on the part of government and conservation organisations in partnership with local communities. Attitudes towards these wild ungulate species have begun to change over the course of 30 years since I first began working in the Himalayan region.

During the summer of 1974, I was on vacation in the Kashmir valley, after a couple years working with Nepal's National Parks Department, when suddenly the northern region of Ladakh opened to foreign visitors. I quickly arranged a trip to Leh and its surroundings. Mr. A.R.Wani, then an officer in the Wildlife Department in Srinagar had just provided me the necessary permits and advice for a short fishing trip I took to lakes above Sonamarg, and he now helped arrange for me to meet with Mr. Desh Paul, the Wildlife Range Officer in Leh. Desh Paul was still overseeing what was essentially a series of trans-Himalayan hunting reserves originally designated by the British, and he took pride in noting when the last argali or ibex were taken by a foreign hunter in the different reserves.

THE AMAZING WORLD OF ELEPHANTS

The Asian Elephant (Elephas maximus)

When you come across an elephant in the wild, or otherwise, you won't need a book to tell you that it is the largest land mammal walking the earth in our era. Their height, weight and strength are almost legendary. Elephants were used by Asian Empires to gain battle superiority. The Kings used them as carriers of nobility. The common man used them for tasks requiring massive feats of strength, like transporting timber.

The elephant was, and to some extent still is, an integral part of this area, especially in India. Elephants have been around for almost 40 million years now. Although, their appearance has evolved to almost unrecognizable proportions in order to adapt to changing environment conditions. The original ancestor of the elephant, known to man, was the Moeritherium. It stood only around 3.5 feet tall and did not have a pronounced trunk like the one we immediately associate with the mention of an elephant. Due to migrations from one area, caused by adverse conditions created there, to other areas with more suitable but different conditions made the ancestors evolve over this long period of time to what we now recognize as the modern day elephant. In this transition stage, the more well known intermediate ascendants after the "little Moeritherium", were the mammoths and the mastodons. The main changes brought about in these species by evolution was in the over all size, the development of the proboscis (or trunk) and the increase in length of the seconds pair of teeth on the upper jaw (tusks). At this point of time, the elephant species is represented by two kinds. The Asian and The African.

The Asian and the African elephants are similar in general appearance but there are a few definite differences. A list of these differences is given below:

	Asian Elephant	**African Elephant**
Height	Average 2 feet shorter (9-10 feet)	Average 2 feet taller (11-12 feet)
Weight	Average 2000 pounds lighter	Average 2000 pounds heavier (can weigh over 8 tons)
Back	Rounded with slight hump in the middle	Have a dip in the middle
Ears	Smaller	Larger and in the shape of the African continent. A pair can weigh up to 110 pounds
Head	Consists of two domes	Consists of a single dome
Tusks	Developed only in males. Female tusks too small to be visible outside	Developed in both male and female
Legs	Shorter and stockier	Longer
Trunk	Has one "finger" at the tip	Has two "fingers" at the tip
Strength	Claimed to be stronger	Not as powerful as the Asian elephant, despite it' s larger size.
General Appearance	Shorter and rounder	Taller and slimmer

The elephant is unique in a couple of characteristics. It is an animal that lives longer in captivity and is also the only domesticated animal, which preferably is taken from the wild for sounder temperament. The average life expectancy of an elephant in the wild is 60 – 70 years, whereas it is around 70 – 80 years in domesticated ones. In the final stages of it's life, the longevity of it's life depends on it's teeth. Elephants go through 6 sets of molars during their life span. The first set of teeth is pushed out by the new set at the age of 2. It is the fourth set at the age of 20 – 25 that is their first set of adult teeth. They get their last set around the age of 60. Once this last set wears out and the elephant can no longer eat, it dies.

Domesticating elephants is an expensive proposition because an elephant in the wild eats approximately 300 pounds of vegetation a day. It's domesticated counter parts are known to eat even more. It consumes a whopping average of 300 liters of water a day, taking in approximately 60 to 80 liters in a single drink.

Another unique and amusing characteristic of elephants is that, despite their huge size and immense weight, their walk can be compared to the dance of a ballerina. Believe it or not but elephants walk on their toes! The heels are elevated with the five digits coming down at a steep angle. Under the heel is a larger cushioning pad of gristle, which other than relieving stress on the bones also allows the animal to be extremely surefooted and silent while walking. The entire "foot" is enclosed in a cylindrical shape and the external toenails are only superficial. It is possible for the number of nails to be different from the number of toes. It can also attain speeds up to 40 kilometers per hour on these feet.

THE ASIATIC LION

(Panthera leopersica)

The Asiatic lion, in India, is now found in only one pocket located in the Gir National Park of the state of Gujarat. There have been efforts made to relocate a small number of these lions to another park to try and ensure better survival chances of this majestic animal. However, these plans are yet to be put into action.

The Asiatic Lion grows to a height of approximately 90cm, with their length ranging from 200 - 280cm. Their long and hard tails grow to an additional length of 60 - 90cm. Lions can weigh in at between 200 - 275kgs. Their main prey species consist of Nilgai, Chital, Sambhar, Goats, Buffaloes and occasionally also other smaller animals. The lions of Gir have made it a habit of killing livestock and sometimes even camels.

Asiatic Lions are built similar to their Afircan cousins but often look smaller due to their smaller manes. Their breeding season is timed along with the onset of winter in the months of October and November. Their gestation period is 3 and a half months, with litters ranging between 3 - 4 in number. The lionesses breed on an average of every two years and are ready to have their first lot of babies around the age of 30 - 36 months. They have a life expectancy ranging between 20 - 30 years.

Just like their African cousins, the Asiatic males too have handsome manes, which are absent in the females. The Asiatic lions tend to have smaller manes than their African cousins. The Asiatic lions however have a bushier overall coat and also have longer tufts of hair at the end of the tail and on the elbow joints. The colouration of the manes vary from lion to lion. The only rare ones are the very dark manes, which according to a research done in Africa are the ones prefered by lionesses!

Lions mostly live in large prides. These prides can sometimes have upto 3 adult males but it is always one that is the dominant leader. The males are known for their laziness and lordship like behaviour. The females do all the hunting, with the males only rarely joining in when the prey is a very large animal like an aggressive buffalo. However, once the kill is made, the males always get the first go at the meal.

The bonding among a family of lions is extremely strong with aunts and sisters helping in the bringing up of all young. Females stay with the pride all through life, whereas the males tend to set out on their own around the age of three. They mostly lead solitary lives then onwards but have been known to also roam territories in pairs and trios. These bachelors are known to be the main threats to the dominant males leading their prides. They are also known to kill cubs to try and get the females into estrus once again. These bachelors are mostly brothers that left a pride together but individual bachelors have been known to team up with other individuals.

Lions, unlike the tiger, hunt in groups. They collectively stalk their prey and have been commonly seen applying strategies that would do any army commander proud. Very often some of the females pinpoint a particular individual prey and chase it in the direction of other lionesses waiting in ambush. The prey is mostly killed by a quick, powerful bite to the spine or with the help of a classic choke grip, with the strong jaws of the lion cutting off air supply to the lungs.

Although history shows the coexistence of lions and tigers, there is no prevalent example of this anywhere in the world at present. Lions do coexist even in the current era with leopards and cheetahs. However, they are extremely territorial and attempt to kill these leopards and cheetahs whenever their paths happen to cross. If the attempt at relocating lions to other parts of India is finally undertaken, it will also answer the question of whether it is possible for two such ferociously territorial and powerful large cats to inhabit the same jungle.

The Asiatic Lion has been declared the most endangered large cat species in the world. Their numbers ranging between 250 - 300, all concentrated in the same area, they are under the constant threat of being wiped out by some deadly epidemic. It is hoped by all conservationists that the governing authorities settle their differences of opinion on the best possible plan and take some action before it's too late to save one of the most magnificent beasts to roam the planet.

SPOTTED DEER / CHITAL

(Axis axis)

The Spotted Deer is the most commonly found member of the deer family in India. It is found across the entire country except in the extreme northern regions. They are mostly seen in large herds of 30 - 50 females with a few stags. However it is not uncommon to see large herds of bachelors numbering upto 100 in areas such as the grasslands of Corbett N.P. They grow to a height of approximately 90 cm at the shoulders and can weigh up to 85 kgs. Their life expectancy ranges from 20 - 30 years. Despite being one of the favourite prey species of predators such as tigers and leopards and only giving birth to a single fawn at a time, their population is quite abundant.

Their diet consists of all kinds of vegetation but grass is the favourite. They also eat the antlers that they shed for their rich nutrients. Spotted deer are very nervous animals and are always on the look out for any approaching danger.

Spotted deer are extremely nervous animals and are always on the alert for a stalking predator. They are often seen under trees housing Langurs for two reasons. Firstly, due to their higher perch, the langurs forewarn them about approaching danger and secondly, the titbits dropped by the Langurs make easy pickings for a meal. The breeding season is not during one part of the year only. It can be in both the summer or winter. However, it is mostly seen during the summers, when males can be seeing throwing their heads back and letting out loud mating calls. Although it comes nowhere close to the volume of the mating call of the Swamp deer. Loud clashing of antlers can also be heard at night during this period.

These deer can be seen, unlike many of their cousins, in dense jungle as well as open grasslands. The main factor behind their location is the presence of water. Due to its alertness and sharp alarm call at seeing a predator, trackers often use this as an indication to pinpoint the location of a predator. However, it is only if the call is repeated many times over that they are taken seriously. A hypersensitive Chital has been seen giving an alarm call due to the falling of a leaf behind it! The alarm calls of Langurs and Sambar deer are taken much more seriously.

Although Chital can be found across the entire country, the widlife parks most famous for their spotted deer populations are Corbett, Kanha, Dudhwa, Bandhavgarh, Mudumalai and Sariska.

HOG DEER

(Axis Porcinus)

The Hog Deer is not as commonly seen as its cousins like the spotted deer or sambar deer. Yet, it does not yet, thankfully, feature in the endangered list. It has a life expectancy of 20 - 30 years. That is, if it manages to escape the attention of predator that long. It grows to a height of approximately 60cm at the shoulders. It is one of the smaller members of the deer family and a fully grown adult mostly weighs in between 20 - 30 kgs. The male members of the species have antlers that can reach lengths of around 30cm, which have a distinctive 3 point formation.

Rutting takes place during the months of autumn and the antlers are shed in spring. They mainly feed on grass, which is why they are mostly spotted in grasslands across the plains and low foothills of northern India. They are, more often than not, seen as solitary animals. However, during the mating season, which ranges around the month of September, they can also be seen in pairs grazing in the alluvial grass plains. The gestation period of these shy creatures is 8 months. One of the ways of identifying them is from the "not long enough" appearance of their legs in comparison to their bodies. If alarmed by you or any suspicious movement or sound, you can also identify them from their long raised tails which are white on the underside. However, this behaviour is not unique to them among the deer species.

Although the Hog Deer can be seen in various parks of India like Corbett, Manas and Dudhwa, the best sighting chances, and that too in pretty large numbers, is at the Kazirangha N.P.

BLUE BULL / NILGAI

(Boselaphus Tragocamelus)

Blue bulls, despite their constant conflict with villagers across almost every part of India, are still present in large numbers. Even though they might be quite commonly seen, a good size male presents quite an awesome picture. They can grow upto a height of 130 - 140 cms at the shoulders and can weigh the same as a decent size horse. They have a life expectancy of 20 -30 years, most of which they prefer to spend in open jungles and scrubby grasslands. They tend to keep away from dense jungles to avoid the obvious danger of lurking predators. Their diet is quite varied and covers almost anything that falls under the term "vegetation".

Nilgai are quite commonly seen all across the central and north India

A decent size male can present quite an awesome sight The male members of this species have horns that grow to a length of 20cm. Adult males are of an excellent dark grey colouration, whereas the females are a lighter fawn. The males also have a coarse dark tuft of hair on the underside of the neck. Another interesting marking is a white patch above each hoof. There is no specific breeding period and they have a gestation period of around 8 - 9 months. Another reason for their growing numbers is that man finds no part of it useful for trade or consumption. It is considered a sacred animal.

Although the Blue bull, which is the largest asiatic antelope, can be seen across many areas of India, there are some reserves where sightings are very abundant. These are Sariska, Sultanpur, Ranthambore and Panna.

INDIAN GAZELLE / CHINKARA

(Gazella Gazella)

This animal is at the other end of the spectrum from the Blue bull. This is the smallest asiatic antelope. It grows to a height of 65. cms and weighs up to a mere 25 kgs. Most males have very short horns, although some grow to a length of 25 - 30 cms. These gazelles are found mostly in open woodlands and scrublands. Their numbers have dwindled to a large extent and they feature on the endangered list. They have a life expectancy of 12 - 15 years, which too is shorter than that of most of its cousins. Their diet consist of all kinds of vegetation. Although their favourite food is lush grass and many types of fruit.

Some of the more well endowed specimens can have horns that grow up to 30 cms.

They have a special characteristic which helps them survive trying times. They can go without water for long periods and can even get sufficient fluids from plants and dew drops. Although mostly seen as solitary animals, they can sometimes be spotted in small groups numbering up to 3 - 4 individuals. They don't have any specific breeding period although it is mostly concentrated around autumn and spring. They have a gestation period of 5 and a half months.

Due to their ability to survive without large sources of water, their habits can vary from one place to another Certain researchers consider the decline in the Chinkara population as the reason behind the Cheetah getting extinct in India. Due to their ability to do without large sources of water, their ideal habitat ranges over varied regions of different topography. The male members of this species is known to be territorial and do not wander large distances like some of their cousins.

They portray a very hyper and nervous appearance due to the constant wagging of their tail or very frequent glances around to check for approaching danger.

The Cinkara gazelle can be seen in various parks across India but the best wildlife parks in which to look out for this animal are Gir, Panna, Ranthambore and the Desert National Park.

SAMBAR DEER

(Cervus Unicolour)

The Sambar Deer is the most widely spread deer species in the world, covering many countries in the Asian continent. It is also one of the larger members of the deer family. Some males are known to weigh up to 300 kgs and can grow to a height ranging from 135 - 150 cms at the shoulders. Their population is large and spread to almost every corner of India. They do not appear on the endangered list. The specimens found in central India appear to be larger than those found in other regions. The male members of this species have antlers that can grow to a length of 90 - 95cms, with a record finding of one that measured 127cms! These animals have a life expectancy ranging between 16 - 20 years. They are the favourite prey species of the tiger. A large sambar can feed a feed tiger for up to 4 days. Unlike the Spotted deer, which shouts an alarm and darts away at the sight of a predator, the sambar tends to alertly watch and keep giving alarm calls until the danger has passed. A reason due to which many of them fall prey for predators.

Sambars are one of the larger members of the deer family and have antlers that can grow to lengths upto 1 meter!

Their own diet consists of vegetation, mainly leaves. They have become less shy of humans due to the protection offered to them at various parks. Yet, they have a history of mainly being nocturnal animals. At Corbett N.P., a sambar deer named Sona by the locals, has adopted the attendants there as the caretakers of itself and its family.

Their breeding period is mainly during the months of November and December. The gestation period is 6 months. The males by this time have shed their antlers. A new pair start growing almost immediately. It is during this period of their life cycles when they are seen less frequently. The males mostly lead solitary lives and are rarely seen associating with each other, except on some occassions during the rutting season. The sambar has extremely sharp senses of hearing and smell.

Its alarm call is taken very seriously, unlike that of the spotted or barking deer, by anyone interested in knowing the whereabouts of a predator. A repeated call is accepted as a definite indicator. Although the Sambar deer is found in almost every corner of India, some of the best parks to sight this animal in are Kanha, Corbett, Ranthambore, Bandhavgarh, Gir, Dudhwa, Manas, Kaziranga and Sariska.

BARKING DEER / KAKAD / MUNTJAC (Muntiacus Muntjac)

This shy and elusive member of the deer family is spread across all the dense jungles of India. It has been named after its call, which bears a striking resemblance to the bark of dog. These animals grow to a height of 50 - 75 cms and weigh 20 -30 kgs. They have a life expectancy of between 20 - 30 years. They mostly live in solitude and are only very rarely seen in numbers exceeding two. Due to their low height and small stature, their main diet consists of grass and fallen fruits. They rarely venture out into open grasslands and are mostly seen feeding near the edge of dense forests. They can also be frequently seen at salt licks like the one shown in the picture below. They are mostly diurnal in habit but it is close to impossible to see them at night due to their dense habitat areas. Their alarm call, unless endlessly repeated, is not taken seriously as an indication of the presence of a predator. They are easily startled by any movement.

A definite identification mark to recognise a barking deer is from the two raised dark ridges on the forehead that extend into its antlers.

A unique trait of the Barking deer is that, unlike other members of the deer family, they possess a pair of antlers as well as overgrown canines known as tushes. Both these are used as weapons in combat but the tushes are used more effectively and frequently.

They are commonly found across the country and have fortunately not yet found their name on the endangered list.

Although the Barking deer is found in almost every corner of India, some of the parks in which we have regularly sighted this animal are Corbett, Bandhavgarh and Sariska.

SWAMP DEER / BARASINGHA

(Cervus Duvauceli)

This handsome member of the deer family is also one that was closest to extinction a few decades ago. Although their numbers have now revived, it is still an endangered species and is spread across central and northern India only in small congregations. It is a medium size member of the deer family. Smaller than the Sambar Deer, it weighs in at approximately

180 kgs once fully grown with a height nearing 130 cms. Their famous antlers, by the time they attain adulthood, develop more than 12 points and attain lengths up to 75 cms. Their life expectancy ranges between 20 - 30 years. Although quite similar in appearance, there are three species with their own unique traits. Their main diet consists of grass which they feed on in the vast grasslands of central and northern India. Some of them have also been seen eating grass from the bed of wet swamps. To do this they have to at times completely submerge their heads and often come out with strands suspended from their antlers.

Swamp Deer are famous for their antlers which have 12 points and give them their Indian name "Barasingha"

They are sociable animals and are often found as large herds grazing in the grasslands of wildlife reserves. The best time to try and spot them is early morning and evening when they emerge to feed.

Swamp Deer are sometimes seen in bachelor groups galloping through grasslands like this "maidan" in Kanha N.P.

The breeding season of the Swamp deer is during the winter months of November and December. They have a gestation period of 6 months and have a litter of mostly just one young. This birth takes place mostly in tall grass where the vulnerable baby is concealed from predators. These animals depend largely on their acute sense of smell to forewarn them about approaching dangers. The species found in central India at reserves such as Kanha N.P. have also adapted to moving into the Sal tree forest from the grasslands.

Their main habitat however still remains dry and wet swampy grasslands. During the breeding season, the males colouration takes on a darker hue and appear a much darker brown when compared to that of the rest of the year. The photograph above was taken during this season. It is also easy to know that you have arrived during the mating season from the unmistakable long drawn out mating calls that carry loudly across the grasslands. These sounded more like "Tarzan yells" to us!

The antlers of the species found in northern India have a shinier and lighter appearance.

Although once found in many forests across the entire country, the Swamp Deer is now mainly confined to National Parks Dudhwa (Northern India), Manas, Kaziranga (North Eastern India), Kanha and Indravati (Central India).

TIGER ON THE ROAD TO EXTINCTION

The death knell has been sounded, and there remains no doubt that the Indian tiger - that awesome and most beautiful of all cats - is now leaving its last few pugmarks on the short road to extinction. There is a sense of helplessness and desperation in all wildlife lovers about the inevitability of the tiger's extinction, fuelled not by the ever-decreasing number of tigers, but by the seeming lack of concern in this area.

Year after year, optimistic tiger census figures, generated by dubious means, are flaunted at us. If these are to be believed, we have a steady tiger population in the country, one that is not showing a downtrend but actually increasing. On the other hand, the various national and international NGOs, vociferously engaged in a last-ditch battle to save the tiger, are giving it no more than ten years of life in the wilds of India. Their contention is based on the rapidly growing incidence of tiger poaching in and around the various national parks in the country, which is also indicated by seizures of tiger skins and parts all over the country.

The recent spurt in tiger deaths near Corbett & Dudhwa Tiger Reserves have all been attributed to poisoning by villagers who, angry at losing livestock to the great predator, resort to lacing the carcass of the cow or buffalo killed by the tiger with massive doses of pesticide. But does this vile, and purely human instinct for revenge, in any way dilute the ghastly fact that the tigers have been poached? The loss that the entire nation suffers when a tiger is killed is surely far greater than the loss of a cow or buffalo of a poor villager. Whereas giving money to the villager can compensate for cattle killed by the feline, can even one dead tiger be replaced? This 'angry villager' syndrome that we seem to be getting trapped in, will be a welcome smoke-screen for the poacher-trader nexus which is thriving brazenly on tiger bones and parts.

It is quite understandable that a free ranging cat like the tiger is bound to come in conflict with humans living in the vicinity of its shrinking habitat, and constantly making inroads or trespassing in core areas. It is also reasonable to give concessions to the people for whom saving the tiger and its habitat is just a government job, and a tough one too. But it is not logical for people who commit lapses in their job, to evade responsibility by hiding behind thin veils.

There was one incident in Dudhwa last year, where the killing of a sub-adult tiger, whose body was found in early stages of decomposition, was explained away by the authorities as death due to intra-specific fight. Billy Arjan Singh cried himself hoarse in meetings and in the newspapers, that it is the rarest of incidents when a tiger will kill another of its kind, but to no avail. To prove his contention that this tiger had been poached by humans, one activist of the Tiger Haven Wildlife Trust came up with a photograph that showed a homespun rope made out of moonjh grass, knotted around the foreleg of that tiger carcass. "The big tiger killed this small one, then tied the rope and wanted to drag him away", was the wry comment of this activist, who absolutely discarded the 'intra-specific fight' theory.

Earlier, there had been more bizzare explanations, when a spurt of killings saw four tiger carcasses floating in the waterbodies in and around Dudhwa. These were attributed to drowning death by suicide, committed by the tigers following rapid increase in their population. It was hard to visualize Indian tigers behaving like lemmings and jumping into the water to commit suicide, when even a layman could understand that a poisoned tiger, whose trachea is choking in spasms, will rush madly towards water in its death throes, as those poor tigers did.

The need of the hour is honesty. On all sides. This virtue, which most of us today don't consider worth passing on to our children, is what the tiger desperately needs, if at all it is to be saved. The forest officials, on their part, must come forward and admit tiger deaths that are caused due to poaching, rather than trying to evade responsibility by evolving ridiculous explanations for them.

Once a tiger death due to poaching is accepted, and the acceptance is backed up by the firm commitment to overcome the lapse, it would become that much easier to go all out in evolving and implementing effective damage-control measures. Otherwise, the forest department's folly will always be evident to others, even though ostrich-like, it may have hidden its head under some convenient, but unlikely, explanation.

On the other hand, there is the growing tendency of non-official conservationists to lay the entire blame on the forest department officials without taking cognizance of their limitations in terms of staff, equipment and commitment. That is the basic reason why these two protagonists, playing out the drama of the tiger's extinction, are constantly at loggerheads. And so, while they keep their horns locked in adversarial contest, vital time is going waste, and nothing concrete is being done. Forest criminals and tiger poachers are having a field day. Trees are being felled to fuel illegal activities of the timber mafia, while tigers are being regularly killed for their skins and bones, which pass through greased palms, and find their way into countries like China, to be sold for astonishing sums in the black market.

The saddest part of this tragic drama is that the focus of the tiger killers has now shifted to Corbett National Park, which was till now considered the best managed tiger reserve, and enjoyed a certain level of immunity from the poacher -- more because of its geographic layout than any human efficiency. But having drastically depleted the tiger population of Dudhwa & Ranthambore, the poachers have now turned towards the relatively healthier tiger population of Corbett Park.

Recent reports in the media have placed the tiger deaths in Corbett at around five in less than a month, even as meetings were being held at the highest levels to mark the Year of the

Tiger, and the 'experts' were congratulating themselves on the formation of some committee or the other to save the tiger. In one fell stroke, the tiger killers had brought to naught all that these committees stand for, and shattered the mythical impregnability of Corbett Tiger Reserve. This, more than anything else, sounds like the death knell.

Much has already been written and said about the inevitability of the tiger's extinction from India, but it still doesn't seem enough to jerk us out of apathy. The recent tiger killings in Corbett, and a glance at hard statistics indicating both optimistic and pessimistic world population estimates, will surely make the reader understand that ten, or even a lesser number of years, are all that remain for the Indian tiger in the wild. The experts don't deny that most tiger census figures are educated guesses, and so it is for the reader to strike that delicate balance between the optimistic and pessimistic estimates of the tiger population.

But whichever way we look at it, if the present downtrend is not arrested, the tiger will fade away into oblivion, and we would be left handling nothing but statistics of what is probably the greatest Indian heritage.

2

WILDLIFE PROTECTION ACT

PRELIMINARY

1. Short title, extent, and commencement, -

(1) This Act may be called the Wildlife (Protection) Act, 1972.

(2) It extends to the whole of India, except the State of Jammu and Kashmir.

(3) It shall come into force in a State or Union Territory to which it extends, on such date as the Central Government may, by notification, appoint, and different dates may be appointed for different provision of this Act or for different States or Union Territories.

2. Definitions - In this Act, unless the context otherwise requires, -

(1) "animal" includes amphibians, birds, mammals, and reptiles, and their young, and also includes, in the cases of birds and reptiles, their eggs,

(2) "animal article" means an article made from any captive animal or wild animal, other than vermin, and includes an article or object in which the whole or any part of such animal [has been used and ivory imported into India and an article made therefrom].

(3) [Omitted 1991.1

(4) "Board means the Wildlife Advisory Board constituted under subsection (1) of Sec. 6;

(5) "captive animal" means any animal, specified in Schedule 1, Schedule II, Schedule III or Schedule IV, which is captured or kept or bred in captivity;

(6) [Omitted 1991.1

(7) "Chief Wildlife Warden" means the person appointed as such under C1. (a) of sub-section (1) of Sec.4;

[(7A) "Circus" means an establishment, whether stationary or mobile where animals are kept or used wholly or mainly for the purpose of performing tricks or manoeuvers;]

(8) "closed area" means the area which is declared under sub-section (1) of Sec.37 to be closed to hunting;

(9) "Collector" means the Chief Officer in charge of the revenue administration of a district;

(10) "commencement of this Act", in relation to –

(a) a State, means commencement of this Act in that State.

(b) any provision of this Act, means commencement of that provision in the concerned State;

(11) "dealer" means any person who carries on the business of buying and selling any captive animal, animal article, trophy, uncurled trophy, [[9]meat or specified plant];

(12) "Director" means the person appointed as Director of Wildlife Preservation under Cl (a). of sub-section (1) of Sec. 3;

[(12A) "Forest Officer" means the Forest Officer appointed under clause (2) of Sec.2 of the Indian Forest Act, 1927 (16 of 1927);]

(13) [Omitted 1991.

(14) "Government property" means property, referred to in sec.39; [or sec. 17H]

(15) "habitat" includes land, water, or vegetation which is the natural home of any wild animal;

(16) "hunting", with its grammatical variations and cognate expressions, includes,

(a) capturing, killing, poisoning, snaring, and trapping or any wild animal and every attempt to do so,

(b) driving any wild animal for any of purposes specified in sub clause

(c) injuring or destroying or taking any part of the body of any such animal, or in the case of wild birds or reptiles, damaging the eggs of such birds or reptiles, or disturbing the eggs or nests of such birds or reptiles;

7) "land" includes canals, creeks, and other water channels, reservoirs, rivers, streams and lakes, whether artificial or natural, [marshes and wetlands and also includes boulders and rocks; [13]].

(18) "licence" means a licence granted under this Act;

(18A) "Live-stock" includes buffaloes, bulls, bullocks, camels, cows, donkeys, goats, horses, mules, pigs, sheep, yak and also includes their young;

(19) "manufacturer" means a manufacturer of animal articles;

(20) "meat" includes blood, bones, sinew, eggs, fat and flesh, whether raw or cooked, of any wild animal other than vermin;

(21) "National Park" means an area declared, whether under sec.35. or sec.38 or deemed, under sub-section (3) of sec.66. to be declared, as a National Park

(22) "notification" means a notification published in the official Gazette;

(23) "permit" means a permit granted under this Act or any rule made thereunder;

(24) "Person" includes a firm;

(25) "prescribed" means prescribed by rules made under this Act;

[15][(25A)"recognised zoo" means a zoo recognised under section 38H;

(25B) "reserve forest" means the forest declared to be reserved by the State Government under sec.20. of the Indian Forest Act, 1927 (16 of 1927);

(26)[16] "sanctuary" means an area declared, whether under sec. [26(A)5]. or sec 38, or deemed, under sub section (3) of Sec.66 to be declared, as a wildlife sanctuary;

(27) "specified plant "means any plant specified in Schedule VI;

(28) "special game" means any animal specified in Sec.II;

(29) "State Government", in relation to a Union Territory, means the Administrator of that Union Territory appointed by the President under Art 239 of the Constitutions ;

(30) "taxidermy", with its grammatical variations and cognate expressions, means the curing, preparation or preservation of trophies;

[(30A) "territorial waters" shall have the same meaning as in Sec.3. of Territorial waters, Continental Shelf, Exclusive Economic Zone and other Maritime Zones Act, 1976 (80 of 1976);

(31) "trophy" means the whole or any part of any captive animal or wild animal, other than vermin, which has been kept or preserved by any means, whether artificial or natural, and includes,

(a) rugs, skins, and specimens of such animals mounted in whole or in part through a process of taxidermy, and

(b) antler, horn, rhinoceros horn, feather, nail, tooth, musk, eggs, and nests;

(32) "uncured trophy" means the whole or any part of any captive animal, other than vermin, which has not undergone a process of taxidermy, and includes a [freshly killed wild animal ambergris, musk and other animal products];

(33) vehicle" means any conveyance used for movement on land, water or air, and includes buffalo, bull, bullock, camel, donkey, elephant, house, and mule;

(34) "vermin" means any wild animal specified in Sch.V;

(35) "weapon" includes ammunition, bows and arrows, explosives, firearms, hooks, knives, nets, poison, snares, traps,

and any instrument or apparatus capable of anaesthetizing, decoying, destroying, injuring or killing an animal;

(36) "wild animal" means any animal found wild in nature and includes any animal specified in Schedule I, Schedule II, Schedule, IV or Schedule V, wherever found;

(37) "wildlife" includes any animal, bees butterflies, crustacean, fish and moths; and aquatic or land vegetation which forms part of any habitat;

(38) "Wildlife Warden" means the person appointed as such under C1. (b) of sub-section (1) of Sec.4;

(39) "zoo" means an establishment, whether stationary or mobile, where captive animals are kept for exhibition to the public but does not include a circus and an establishment of a licenced dealer in captive animals.

The Act has been made applicable in various States and Union Territories as under:

1. Andaman and Nicobar, w.e.f. Ist July 1973, vide G.S.R.332 (E), dated Ist July, 1973.

2. Andhra Pradesh, w.e.f. Ist July 1973, vide G.S.R.371 (E), dated Ist August, 1973.

3. Arunachal Pradesh, w.e.f. 15th May 1973.

4. Assam, w.e.f. 25th January 1977, vide G.S.R.33 (E), dated 25th January, 1977.

5. Bihar, w.e.f. Ist February 1973, vide G.S.R.40 (E), dated Ist February, 1973.

6. Dadra and Nagar Haveli, w.e.f. Ist September 1973, vide G.S.R.441(E), dated Ist September, 1973.

7. Delhi, w.e.f. Ist June 1973,

8. Gao Daman and Diu, w.e.f. 16 July 1973, vide G.S.R.629 (E), dated 16th July, 1973.

9. Gujarat, w.e.f. Ist February 1973, vide G.S.R.62 (E), dated Ist February, 1973.

10. Haryana, w.e.f. 12th March July 1973, vide G.S.R.63 (E), dated 12 March, 1973.

11. Himachal Pradesh, w.e.f. 2nd April ly 1973, vide G.S.R.190 (E), dated 2nd April, 1973.

12. Kerala, w.e.f. Ist June 1973, vide G.S.R.293 (E), dated Ist June, 1973.

13. Lakshdweep, w.e.f. Ist November 1973, vide G.S.R.481(E), dated Ist November, 1973.

14. Madhya Pradesh, w.e.f. 25 January 1973, vide G.S.R.28 (E), dated 25 January, 1973.

15. Maharashtra, w.e.f. Ist June 1973, vide G.S.R.296 (E), dated Ist June, 1973.

16. Manipur, w.e.f. 15th May 1973, vide G.S.R.269 (E), dated 15th May, 1973.

17. Meghalaya, w.e.f. Ist April 1977

18. Mizoram, w.e.f. Ist October 1974, vide G.S.R.407 (E), dated Ist October, 1974.

19. Mysore, w.e.f. Ist July 1973, vide G.S.R.355 (E), dated Ist July, 1973.

20. Nagaland, w.e.f. 18th December 1981, vide G.S.R.668 (E), dated 18th December

21. Orissa, w.e.f. 14th August 1974, vide G.S.R.364 (E), dated 14th August, 1974.

22. Punjab, w.e.f. Ist April 1975.

23. Rajasthan, w.e.f. Ist September 1973, vide G.S.R.410 (E), dated Ist September, 1973.

24. Sikkim, w.e.f. Ist May 1976, vide G.S.R.311 (E), dated Ist May, 1976.

25. Tamil Nadu, w.e.f. Ist January, 1974.

26. Tripura, w.e.f. 2nd October 1973, vide GS.R.465 (E), dated 2nd October, 1973.

27. Union Territory of Chandigarh, w.e.f. 2nd December 1974, vide G.S.R.674 (E), dated 2nd December, 1974.

28. Union Territory of Pondicherry, w.e.f. Ist March 1975, vide G.S.R.62 (E), dated Ist March, 1975

29. Uttar Pradesh, w.e.f. Ist February 1973.

30. West Bengal, w.e.f. Ist May 1973, vide G.S.R.224 (E), dated Ist May, 1973.

AUTHORITIES TO BE APPOINTED OR CONSTITUTED UNDER THIS ACT

3. Appointment of Director another officers.–

(1) The Central Government may, for the purposes of this Act appoint

(a) a Director of Wildlife Preservation;

(b) Assistant Directors of Wildlife Preservation; and

(c) such other officers and employees as may be necessary.

(2) In the performance of his duties and exercise of his powers by or under this Act, the Director shall be subject to such general or special directions, as the Central Government may, from time to time, give.

(3) The Assistant, Directors of Wildlife Preservation and other officers and employees appointed under this section shall be subordinate to the Director.

4. Appointment of Chief Wildlife Warden and other officers–

(1) The Rate Government may, for the purposes of this Act, appoint –

(a) a Chief Wildlife Warden;

(b) Wildlife Wardens;

(bb) One Honorary Wildlife Ward in each district; and

(c) such other officers and employees as may be necessary.

(2) In the performance of his duties and exercise of his powers by or under this Act, the Chief Wildlife Warden shall be subject to such general or special directions, as the State Government may, from time to time, give.

(3) [The Wildlife Warden, the Honorary Wildlife Warden] and other officers and employees appointed under this section shall be subordinate to the Chief Wildlife Warden.

5. Power of delegate –

(1) The Director may, with the previous approval of the Central Government, by order in writing delegate all or any of his powers and duties under this Act to any officer subordinate to him subject to such conditions, if any, as may be specified in the order.

(2) The Chief Wildlife Warden may, with the previous approval of the State Government by order in writing, delegate all or any of his powers and duties under this Act, except those under Cl.(a) of sub-section (1) of Sec.11, to any officer subordinate to him subject to such conditions, if any, be specified in the order.

(3) Subject to any general or special, direction given or condition imposed by the Director or the Chief Wildlife Warden, any person, authorised by the Director or the Chief Wildlife Warden to exercise any, powers, may exercise those powers in the same manner and to the same effect as if they had been conferred on that person directly by this Act and not by way of delegation.

6. Constitution of the Wildlife Advisory Board.–

(1) The State Government, or, in the case of a Union Territory, the Administrator, shall, as soon as may be after the commencement of this Act, constitute a Wildlife Advisory Board consisting of the following member, namely:

(a) the Minister in charge of Forest in the State or Union Territory, or, if there is no such Minister, the Chief Secretary to the State Government, or, as the case may be, the Chief Secretary to the Government or the Union Territory, who shall be the Chairman;

(b) two members of the State Legislature or, in the case of a Union Territory havinga Legislature, two members of the legislature of the Union Territory, as the case may be;

(c) Secretary to the State Government, or the Government of the Union Territory, in charges of Forests;

(d) The Forest Officer in charge of the State Forest Department, by whatever designation called, ex-officio; [4]]

(e) an officer to be nominated by the Director;

(f) Chief Wildlife Warden, ex-officio;

(g) Officers of the State Forest Government not exceeding five;

(h) such other person, not exceeding ten, who, in the opinion of the State Government, are interested in the protection of Wildlife, including the representatives of tribals not exceeding three.

(1A) The State Government may appoint a Vice-Chairman of the Board from amongst the members referred to in clauses (b) and (h) of sub-section (1).

(2) The State Government shall appoint [the forest Officer incharge of the Forest Department or Chief Wildlife Warden as the Secretary to the Board.

(3) The term of office of the members of the Board referred to in C1.(g) of sub-section (1) and the manner of filling the vacancies among them shall be such as may be prescribed. [8]

(4) The members shall be entitled to receive such allowances in respect of expenses incurred in the performance of their duties as the State Government may prescribed.

7. Procedure to be followed by the Board. –

(1) The Board shall meet at least twice a year at such place as the State Government may direct.

(2) The Board shall regulate its own procedure (including the quorum).

(3) No act or proceeding of the Board shall be invalid merely by reason of the existence of any vacancy therein or any defect in the constitution thereof or any irregularity in the procedure of the Board affecting the merits of the case.

8. Duties of the Wildlife Advisory Board. – It shall be the duty of the Wildlife Advisory Board to advise the State Government,–

(a) In the selection of areas to be declared as Sanctuaries, National Parks, and Closed Areas and the administration thereof;

(b) mulation of the policy of protection and conservation of Wil nd specified plants;

(c) any matter relating to any schedule;

(cc) in relation to the measures to be taken for harmonizing the needs of the tribals and other dwellers of the forest with the protection and conservation of wildlife; and

(d) in any matter that may be referred to it by the State Government.

HUNTING OF WILD ANIMALS

(9) Prohibition of Hunting.

No person shall hunt any wild animal specified in Schedule, I, II, III and IV except as provided under section 11 and section 12.

10. (Omitted 199)

11. Hunting of Wild animals to be permitted in certain cases.–

(1) Notwithstanding anything contained in any other law for the time being in force and subject to the provisions of Chapter IV –

(a) the Chief Wildlife Warden may, if he is satisfied that any wild animal specified in Sch. 1 has become dangerous to human life or is so disabled or diseased as to be beyond recovery, by order in writing and stating the reasons therefor, permit any person to hunt such animal or cause animal to be hunted;

(b) the Chief Wildlife Warden or the authorised officer may, if he is satisfied that any wild animal specified in Sch. II Sch, III or Sch. IV has become dangerous to human life or to property (including standing crops on any land) or is so disabled or diseased as to be beyond recovery, by order in writing and stating the reasons therefor, permit any person to hunt such animal or cause such animal to be hunted.

(2) The killing or wounding in good faith of any wild animal in defence of oneself or of any other person shall not be an

offence; Provided that nothing in this sub-section shall exonerate any person who, when such defence becomes necessary, was committing any act in contravention of any provision of this Act or any rule or order made thereunder.

(3) Any wild animal killed or wounded in defence of any person shall be Government property.

12. Grant of permit for special purposes. - Notwithstanding anything contained elsewhere in this Act, it shall be lawful for the Chief Wildlife Warden, to grant a permit, by an order in writing stating the reasons therefor, to any person, on payment of such fee as may be prescribed, which shall entitle the holder of such permit to hunt, subject to such conditions as may be specified therein, any wild animal specified in such permit, for the purpose of, –

(a) education;

(b) scientific research;

(bb) scientific management;

Explanation:- for the purposes of Cl. (bb), the expression, "scientific management" means –

(i) translocation of any wild animal to an alternative suitable habitat; or

(ii) population management of wildlife, without killing or poisoning or destroying any wild animals.

(c) Collection of specimens

(i) for recognised zoos subject to the permission under section 38-1 or

(ii) for museums and similar institutions;

(d) derivation, collection or preparation of snake-venom for the manufacture of life saving drugs. Provided that no such permit shall be granted:

(a) in respect of any wild animal specified in Sch.L except with the previous permission of the Central Government, and

(b) in respect of any other wild animal, except with the previous permission of the State Government.

(13) "Suspension or cancellation of licence. -

The Chief Wildlife Warden or the authorised officer may, subject to any general or special order of this State Government, for good and sufficient reasons, to be recorded in writing, suspend or cancel any licence granted under this Chapter. (OMITTED 1991)

Provided that no such suspensions or cancellation shall be made except after giving the holder of the licence a reasonable opportunity of being heard".

PROTECTION OF SPECIFIED PLANTS

17A. Prohibition of picking, uprooting, etc., of specified plants. – Save, as otherwise provided in this Chapter, no person shall–

(a) willfully pick, uproot, damage destroy, acquire or collect any specified plant from any forest land and area specified, by notification, by the Central Government,

(b) possess, sell, other for sale, or transfer by way of gift or otherwise, or transport any specified plant, whether alive or dead, or part or derivative thereof :

Provided that nothing in this section shall prevent a member of a scheduled tribe, subject to the provisions of Chapter IV, from picking, collecting or possessing in the district he resides any specified plant or part or derivative thereof for his bonafide personal use.

17B. Grant of permit for special purpose–. The Chief Wild Life Warden may with the previous permission of the State Government, grant to any person a permit to pick, uproot, acquire or collect from a forest land or the area specified under section 17A or transport, subject to such conditions as may be specified therein, any specified plant for the purpose of

(a) education;

(b) scientific research.,

(c) collection, preservation and display in a herbarium of any scientific institutions; or

(d) propagation by a person or an institution approved by the Central Government in this regard.

17C. Cultivation of specified plants without licence prohibited–

(1) No person shall cultivate a specified plant except under, and in accordance with a licence granted by the Chief Wild Life Warden or any other officer authorised by the State Government in this behalf : Provided that nothing in this section shall prevent a person, who, immediately before the commencement of the Wild Life (Protection)

Amendment Act, 1991, was cultivating a specified plant, from carrying on such cultivation for a period of six months from such commencement, or where he has made an application within that period for the grant of a licence to him, until the licence is granted to him, or he is informed in writing that a licence cannot be granted to him.

(2) Every licence granted under this section shall specify the area in which and the conditions, if any, subject to which the licensee shall cultivate a specified plant.

17D. Dealing in specified plants without licence prohibited-

(1) No person shall, except under and in accordance with a licence granted by the Chief Wild Life Warden or any other officer authorised by the State Government in this behalf, commence or carry on business or occupation as a dealer in a specified plant or part or derivative thereof :

Provided that nothing in this section shall prevent a person, who, immediately before that commencement of the Wild Life (Protection) Amendment Act, 1991, was carrying on such business or occupation for a period of sixty days from such commencement, or where he has made an application within that period for the grant of a licence to him, until the licence is granted to him or he is informed in writing that a licence cannot be granted to him.

(2) Every licence granted under this section shall specify the premises in which and the conditions, if any, subject to which the licensee shall carry on his business.

17E. Declaration of stock.–

(1) Every person cultivating, or dealing in, a specified plant or part or derivative thereof shall, within thirty days, from the date of commencement of the Wild Life (Protection) Amendment Act, 1991, declare to the Chief Wild Warden or any other officer authorised by the State Government in this behalf, his stocks of such plants and part or derivative thereof, as the case may be, on the date of such commencement.

(2) The provisions of sub-section (3) to (8) (both inclusive) of section 44, section 45, section 46 and section 47 shall, as far as may be, apply in relation to an application and a licence referred to in section 17C and section 17D as they apply in relation to the liecence or business in animals or animal articles.

17F. Possession, etc., of plants by licensee. -

No licensee under this chapter shall

(a) keep in his control or possession

(i) any specified plant, or part or derivative thereof in respect of which a declaration under the provisions of section 17E has to be made, but has not been made;

(ii) any specified plant, or part or derivative thereof which has not been lawfully acquired under the provisions of this Act or any rule, or order made thereunder;

(b) (i) pick, uproot, collect or acquire any specified plant, or

(ii) acquire, receive, keep in his control, custody or possession, or sell, offer for sale or transport, any specified plant or part or derivative thereof, except in accordance with the conditions subject to which the licence has been granted and such rules as may be made under this Act.

17G. Purchase, etc., of specified plants.– No person shall purchase, receive or acquire any specified plant or part or derivative thereof otherwise than from a licenced deader: Provided that nothing in this section shall apply to any person referred to in section of 17B.

17H. Plants to be Government property. –

(1) Every specified plant or part or derivative thereof, in respect of which any offence against this Act or any rule or order made thereunder has been committed, shall be the property of the State Government, and, where such plant or part or derivative thereof has been collected or acquired from a sanctuary or National Park declared by the Central Government, such plant or part or derivative thereof shall be the property of the Central Government.

(2) The provisions of sub-sections (2) and (3) of section 39 shall, as far as may be, apply in relation to the specified plant or part or derivative thereof as they apply in relation to wild animals and articles referred to in sub-section (1) of that section.]

SANCTUARIES, NATIONAL PARK, AND CLOSED AREAS SANCTUARIES

18. Declaration of Sanctuary.–

(l) The State Government may, by notification, declare its intention to constitute any area other than area comprised with any reserve forest or the territorial waters as a sanctuary if it considers that such area is of adequate ecological, faunal, floral, geomorphological, natural. or zoological significance, for the purpose of protecting, propagating or developing wildlife or its environment.

(2) The notification referred to in sub-section (1) shall specify, as nearly as possible, the situation and limits of such area.

Explanation. - For the purposes of the this section, it shall be sufficient to describe the area by roads, rivers, ridges, or other well-known or readily intelligible boundaries

19. Collector to determine rights.–

When a notification has been issued under Sec.18,] the collector shall inquire into, and determine the existence, nature and extent of the rights of any person in or over the land comprised within the limits of the sanctuary.

20. Bar of accrual of rights.–

After the issue of a notification under Sec." 18, no right shall be acquired in, or over the land comprised within the limits of the area specified in such notification, except by succession, testamentary or intestate.

21. Proclamation by Collector. –

When a notification has been issued under Sec.18 the Collector shall publish in the regional language in every town and village in or in the neighborhood of the area comprised therein, a progamation:

(a) specifying, as nearly as possible, the situation and the limits of the sanctuary; and

(b) requiring any person, claiming any right mentioned in Sec. 19, to prefer before the collector" within two months from the date of such proclamation, a written claim in the prescribed form specifying the nature and extent of such right, with necessary details and the amount and particulars of the compensation, if any, claimed in respect thereof.

22. Inquiry by Collector. –

The Collector shall, after service of the prescribed notice upon the claimant, expeditiously inquire into

(a) the claim preferred before him under Cl. (b) of Sec.21, and

(b) the existence of any right mentioned in Sec.19 and not claimed under Cl.(b) of Sec.21, so far as the same may be ascertainable from the records of the State Goven-iments and the evidence of any person acquainted with the same.

23. Powers of Collector. –

For the purpose of such inquiry, the Collector may exercise the following powers, namely

(a) the power to enter in or upon any land and to survey, demarcate, and make a map of the same or to authorise any other officer to do so;

(b) the same powers as are vested in a civil court for the trial of suits.

24. Acquisition of rights. –

(1) In the case of a claim to a right in or over any land referred to in Sec.19, the Collector shall pass an order admitting or rejecting the same in whole or in part.

(2) If such claim is admitted in whole or in part, the Collector may either

(a) exclude such land from the limits of the proposed sanctuary, or

(b) proceed to acquire such land or rights, except where by an agreement between the owner of such land or the holder of rights and the Government the owner or holder of such rights has agreed to surrender his rights to the Government, in or over such land, and payment of such compensation, as is provided in the Land Acquisition Act, 1894 (1 of 1894)

(c) allow, in consultation with the Chief Wildlife Warden, the continuance of any right of any person in, or over any land within the limits of the sanctuary.

25. Acqitisition proceedings. –

(1) For the purpose of acquiring such land, or rights in or over such land,

(a) the Collector shall be deemed to be a Collector, proceeding under the Land Acquisition Act, 1894 (1 tó 1894):

(b) the claimant shall be deemed to be a person interested and appearing before him in pursuance of a notice given under sec.9 of that Act.

(c) the provisions of the sections preceding Sec.9 of that Act shall be deemed to have been complied with;

(d) where the claimant does not accept the award made in his favour in the matter of compensation, he shall be deemed, within the meaning of Sec.18 of that Act, to be a person interested who has not accepted the award, and shall be entitled to proceed to claim relief, against the award under the provision of Part III of that Act;

(e) the Collector, with the consent of the claimant, or the Court, with the consent of both the parties, may award compensation in land or money or partly in land and partly in money, and

(f) in the case of the stoppage of a public way or a common pasture, the Collector may, with the previous sanction of the State Government provide for an alternative public way or common pasture, as far as may be practicable or convenient.

(2) The acquisition under this Act of any land or interest therein shall be deemed to be acquisition for a public purpose.

26. Delegation of Collector's powers. –

The State Government may, by general or special order, direct that the powers exercisable or the functions to be performed by the Collector under Sec. 19 to 25 (both inclusive) may be exercised and performed by such other officer as may be specified in the order.

(26A) Declaration of area as Sanctuary. –

(1) When –

(a) a notification has been issued under sec.18 and the period for preferring claim has elapsed, and all claims, if any, made in relation to any land in an area intended to be declared as a sanctuary, have been disposed of by the State Government; or

(b) any area comprised within any reserve forest or any part of the territorial waters, which is considered by the State Government to be of adequate ecological, faunal, geomorphological, natural or zoological significance for the purpose of protecting, propagating or developing wildlife or its environment, is to be included in a sanctuary, the State Government shall issue a notification specifying the limits of the area which shall be comprised within the sanctuary and declare that the said area shall be sanctuary on and from such date as may be specified in the notification.

Provided that where any part of the territorial waters is to be so included, prior concurrence of the Central Government shall be obtained by the State Government. Provided further that the limits of the area of the territorial waters to be included in the sanctuary shall be determined in consultation with the Chief Naval Hydrographer of the Central Government and after taking adequate measures to protect the occupational interests of the local fishermen.

(2) Notwithstanding anything contained in sub-section (1), the right of innocent passage of any vessel or boat through the territorial water shall not be affected by the notification issued under sub-section (1).

(3) No alteration of the boundaries of a sanctuary shall be made except on a resolution passed by the Legislation of the State.]

27. Restriction on entry in sanctuary. –

(1) No person other than,

(a) a public servant on duty;

(b) a person who has been permitted by the Chief Wildlife Warden or the authorised officer to reside within the limits of the sanctuary;

(c) a person who has any right over immovable property within the limits of the sanctuary;

(d) a person passing through the sanctuary along a public highway, and

(e) the dependents of the person referred to in CI. (a), (b) or (c) shall enter or reside in the sanctuary, except under and in accordance with the conditions of a permit granted under section 28.

(2) Every person shall, so long as he resides in the sanctuary, be bound

(a) to prevent the commission, in the sanctuary, or an offence against this Act;

(b) where there is reason to believe that any such offence against this Act has been committed in such sanctuary, to help in discovering and arresting the offender;

(c) to report the death of any wild animal and to safeguard its remains until the Chief Wildlife Warden or the authorised officer takes charge thereof;

(d) to extinguish any fire in such sanctuary of which he has knowledge or information and to prevent from spreading by any lawful means in his power , any fire within the vicinity of such sanctuary of which he has knowledge or information; and

(e) to assist any forest officer, Chief Wildlife Warden, Wildlife Warden or police officer demanding his aid for preventing the commission of any offence against this Act or in the investigation of any such offence.

(3) No person shall, with intent to cause damage to any boundary-mark of a sanctuary or to cause any wrongful gain as defined in the Indian Penal Code (45 of 1860), alter, destroy, move, or deface such boundary-mark.

(4) No person shall tease or molest any wild animal or litter the grounds or sanctuary.

28. Grant of permit. –

(1) The Chief Wildlife Warden may, on application, grant to any person a permit to enter or reside in a sanctuary for all or any of the following purposes, namely:

(a) investigation or study of wildlife and purposes ancillary or incidental thereto;

(b) photography;

(c) scientific research;

(d) tourism;

(e) transaction of lawful business with any person residing in the sanctuary.

(2) A permit to enter or reside in a sanctuary shall be issued subject to such conditions and on payment of such fee as may be prescribed.

(29) Destruction, etc., in a sanctuary prohibited without a permit. –

No person shall destroy, exploit or remove any wildlife from a sanctuary or destroy or damage the habitat of any wild animal or deprive any wild animal or its habitat within such sanctuary except under and in accordance with a permit granted by the Chief Wildlife Warden and no such permit shall be granted unless the State Government being satisfied that such destruction, exploitation or removal of wildlife from the sanctuary is necessary for the improvement and better management of wildlife therein authorises the issue of such permit.

Explanation: For the purposes of this section, grazing or movement of livestock permitted under clause (d) of Sec.33 shall not be deemed to be an act prohibited under this section.]

30. Causing fire prohibited. –

No person shall set fire to a sanctuary, or kindle any fire, or leave any fire burning, in a sanctuary, in such manner as to endanger such sanctuary.

31 Prohibition of entry into sanctuary with weapon. –

No person shall enter a sanctuary with any weapon except with the previous permission in writing of the Chief Wildlife Warden or the authorised officer.

32. Ban on use of injurious substances. –

No person shall use in a sanctuary, chemicals, explosives or any other substances which may cause injury to, or endanger, any wildlife in such sanctuary.

33. Control of sanctuaries. –

The Chief Wildlife Warden shall be the authority who shall control, manage and maintain all sanctuaries and for that purpose, within the limits of any sanctuary,

(a) may construct such roads, bridges, buildings, fences or barrier gates, and carry out such other works as he may consider necessary for the purposes of such sanctuary;

(b) shall take such steps as will ensure the security of wild animals in the sanctuary and the preservation of the sanctuary and wild animals, therein;

(c) may take such measures, in the interests of wildlife, as he may consider necessary for the improvement of any habitat.

(d) may regulate, control or prohibit, in keeping with the interests of wildlife, the grazing or movement of [livestock].

(e) [omitted 19911

33A. Immunisation of livestock. –

(1) The Chief Wildlife Warden shall take such measures in such manner as may be prescribed, for immunisation against communicable diseases of the livestock kept in or within five kilometres of a sanctuary.

(2) No person shall take, or cause to be taken or graze, any livestock in a sanctuary without getting it immunised.]

34. Registration of certain persons in possession of arms.

(1) Within three months from the declaration of any area as a sanctuary, every person residing in or within ten kilometres of any such sanctuary and holding a licence granted under the Arms Act, 1959 (54 of 1959), for the possession of arms or exempted from the provisions of that Act and possessing arms, shall apply in such form, on payment of such fee, and within such time as may be prescribed, to the Chief Wildlife Warden or the authorised officer, for the registration of his name.

(2) On receipt of an application under sub-section (1), the Chief Wildlife Warden or the authorised officer shall register the name of the applicant in subject manner as may be prescribed.

(3) No new licences under the Arms Act, 1959 (54 of 1959), shall be granted within a radius of ten kilometres of a sanctuary without the prior concurrence of the Chief Wildlife Warden.

NATIONAL PARKS

35. Declaration of National Parks. –

(1) Whenever it appears to the State Government that an area, whether within a sanctuary or not, is, by reason of its ecological, faunal, floral, geomorphological, or zoological association or importance, needed to be constituted as a National Park for the purpose of protectin& propagating or developing

wildlife therein or its environment, it may, by notificLtion, declare its intention to constitute such area as a National Park.

(1) Provided that where any part of the territorial waters is proposed to be included in such National Park, the provisions of Sec.26A shall, as far as may be, apply in relation to the declaration of a National Park as they apply in relation to the declaration of a sanctuary.

(2) The notification referred to in sub-section (1) shall define the limits of the area which is intended to be declared as a National Park.

(3) Where any area is intended to be declared as a National Park, the provisions of Sec. [[12]19 to 26-A (both inclusive except clause (c) of sub-section (2) of section 24)] shall, as far as may be, apply to the investigation and determination of claims and extinguishment of rights, in relation to any land in such area as they apply to the said matters in relation to any land in a sanctuary.

(4) When the following events have occurred, namely

(a) the period for preferring claims has elapsed, and all claims, if any, made in relation to any land in an area intended to be declared as a National Park, have been disposed of by the State Government, and

(b) all rights in respect of lands proposed to be included in the National Park have become vested in the State Government the State Government shall publish a notification specifying the limits of the area which shall be comprised within the National Park and declare that the said area shall be a National Park on and from such date as may be specified in the notification.

(5) No alteration of the boundaries of a National Park shall be made except on a resolution passed by the Legislature of the State.

(6) No person shall, destroy, exploit, or remove any wildlife from a National Park or destroy or damage the habitat or any wild animal or deprive any wild animal or its habitat within such National Park except under and in accordance with a

permit granted by the Chief Wildlife Warden and no such permit shall be granted unless the State Government, being satisfied that such destruction, exploitation, or removal of wildlife from the National Park is necessary for the improvement and better management of wildlife therein, authorises the issue of such permit.

(7) No grazing of any [livestock[13]] shall be permitted in a National Park and no livestock shall be allowed to enter except where such [livestock] is used as a vehicle by a person authorised to enter such National Park.

(8) The provisions of secs. 27 and 28, secs.30 to 32 (both inclusive), and CIS, (a), (b) and (c) of [Sec.33, 33A[14]] and sec.34 shall, as far as may be, apply in realtion to a National Park as they apply in relation to a sanctuary.

CLOSED AREA

37. Declaration of closed area. –

(1) The State Government may, by notification, declare any area closed to hunting for such period as may be specified in the notification.

(2) No hunting of any wild animal shall be permitted in a closed area during the period specified in the notification referred to in sub-section(1).

Sanctuaries or National Park Declared by Central Govt.

38. Power of Central Government to declare areas as Sanctuaries or National Park, -

(1) Where the State Government leases or otherwise transfers any area under its control, not being an area within a Sanctuary, to the Central Government the Central Government may, if it is satisfied that the conditions specified in sec.18 are fulfilled in relation to the area so transferred to it, declare such area, by notification, to be a sanctuary and the provisions of sec 18 to 35 (both inclusive), 54 and 55 shall apply in relation to such sanctuary as they apply in relation to a sanctuary declared by the State Government.

(2) The Central Government may, if it is satisfied that the conditions specified in sec.35 are fulfilled in relation to any area referred to in sub-section (1), whether or not such area has been declared, to be a sanctuary by the Central Government, or the State Government, declare such area, by notification, to be a National Park and the provisions of secs.35. 54 and 55 shall apply to such National Park as they apply in relation to a National Park declared by the State Government.

(3) In relation to a sanctuary or National Park declared by the Central Government, the powers and duties of the Chief Wildlife Warden under the section referred to in sub-section (1) and (2). shall be exercised and discharged by the Director or by such other officer as may be authorised by the Director in this behalf and references in the sections aforesaid to the State Government, shall be construed as reference to the Central Government and reference therein to the Legislation of the State shall be construed as a reference to Parliament.

CENTRAL ZOO AUTHORITY AND RECOGNITION OF ZOOS

38A. Constitution of Central Authority. -

(1) The Central Government shall constitute a body to be known as the Central Zoo Authority (hereinafter in this Chapter referred to as the Authority), to exercise the powers conferred on, and to perform the functions assigned to it under this Act.

(2) The Authority shall consist of

(a) Chairperson;

(b) such number of members not exceeding ten; and

(c) Member - Secretary; to be appointed by the Central Government.

38B. Term of office and conditions of service of chairperson and members etc.–

(1) The chairperson and every member shall hold office for such period, not exceeding three years, as may be specified by the Central Government.

(2) The chairperson or a member may, by writing under this hand, addressed to the Central Government, resign from the office of chairperson or, as the case may be, of the Member.

(3) The Central Government shall remove a person from the office of chairperson or member referred to in sub-section (2) if that person.

(a) becomes an undischarged insolvent,

(b) gets convicted and sentenced to imprisonment for an offence which, in the opinion of the Central Government, involves moral turpitude;

(c) becomes of unsound mind and stands so declared by a competent court;

(d) refuses to act or becomes incapable of acting;

(e) is, without obtaining leave or absence from the authority, absent from three consecutive meetings of the Authority; or

(f) in the opinion of the Central Government has so abused the position of chairperson or member as to render that person's continuance in office detrimental to the public interest:

Provided that no person shall be removed under this clause unless that person has been given a reasonable opportunity of being heard in the matter.

(4) A vacancy caused under sub-section (2) or otherwise shall be filled by fresh appointment.

(5) The salaries and allowances and other conditions of appointment of chairperson, members and Member-Secretary of the Authority shall be such as may be prescribed.

(6) The Authority shall, with the previous sanction of the Central Government, employ such officer and other employees as it deems necessary to carry out the purposes of the Authority.

(7) The terms and conditions of service of the officers and other employees of the authority shall be such as may be prescribed.

(8) No act or proceeding of the Authority shall be questioned or shall be invalid on the ground merely of the existence of any vacancies or defect in the constitution of the Authority.

38C. Functions of the Authority. – The Authority shall perform the following functions, namely:

(a) specify the minimum standards for housing, unkeep and veterinary care of the animals kept in a zoo;

(b) evaluate and assess the functioning of zoos with respect to the standards or the norms as may be prescribed;

(c) recognise or derecongnize zoos;

(d) identify endangered species of wild animals for purposes of captive breeding and assigning responsibility in this regard to a zoo;

(e) co-ordinate the acquisition, exchange and loaning of animals for breeding purposes;

(f) ensure maintenance of stud-books of endangered species of wild animals bred in captivity;

(g) identify priorities and themes with regard to display of captive animals in a zoo;

(h) co-ordinate training of zoo personnel in India and outside India;

(i) co-ordinate research in captive breeding and educational programmes for the purposes of zoos;

(j) provide technical and other assistance to zoos for their proper management and development on scientific lines;

(k) perform such other functions as may be necessary to carry out the purposes of this Act with regard to zoos.

38D. Procedure to be regulated by the Authority. –

(1) The Authority shall meet as and when necessary and shall meet at such time and place as the chairperson may think fit.

(2) The Authority shall regulate its own procedure.

(3) All orders and decisions of the Authority shall be authenticated by the Member-Secretary or any other officer of the Authority duly authorised by the Member-Secretary in this behalf.

38E. Grants and loans to Authority and constitution of found.–

(1) The Central Government may, after due appropriation made by Parliament by law in this behalf, make to the Authority grants and loans of such sums of money as that Government may consider necessary.

(2) There shall be constituted a Fund to be called the Central Zoo Authority Fund and there shall be credited thereto any grants and loans made to the Authority by the Central Government all fees and charges received by the Authority under this Act and all sums received by the Authority from such other sources as may be decided upon by the Central Government.

(3) The Fund referred to in sub-section (2) shall be applied for meting salary, allowances and other remuneration of the members, officer and other employees of the Authority and the expenses of the Authority in the discharge of its functions under this Chapter and expenses on objects and for purposes authorised by this Act.

(4) The Authority shall maintain proper accounts and other relevant records and prepare an annual statement of accounts in such form as may be prescribed by the Central Government in consultation with the Comptroller and Auditor General of India.

(5) The accounts of the Authority shall be audited by the Comptroller and Auditor-General at such intervals as may be specified by him and any expenditure incurred in connection with such audit shall be payable by the Authority to the Comptroller and Auditor-General.

(6) The Comptroller and Auditor-General and any person appointed by him in connection with the audit of the accounts of the Authority under this Act shall have the same rights and

privileges and the authority in connection with such audit as the Comptroller and Auditor-General generally has in connection with the audit of the Government accounts and, in particular, shall have the right to demand the production of books, accounts, connected vouchers and other documents and papers and to inspect any of the offices of the Authority.

(7) The accounts of the Authority as certified by the Comptroller and Auditor-General or any other person appointed by him in this behalf together with the audit report thereon, shall be forwarded annually to the Central Government by the Authority.

38F. Annual report. -

The Authority shall prepare in such form and at such time, for each financial year, as may be prescribed, its annual report, giving a full account of its activities during the previous financial year and forward a copy thereof to the Central Government.

38G. Annual report and audit report to be laid before Parliament. -

The Central Government shall cause the annual report together with a memorandum of action taken on the recommendations contained therein, in so far as they relate to the Central Government, and the reasons for the non-acceptance, if any, of any of such recommendations and the audit report to be laid as soon as may be after the reports are received before each House of Parliament.

38H. Recognition of Zoos. –

(1) No zoo shall be operated without being recognised by the Authority. Provided that a zoo being operated immediately before the date of commencement of the Wild Life (Protection) Amendment Act, 1991 may continue to operate without being recognised for a period of [eighteen months from the date of such commencement and[3]] if the application seeking recognition is made within that period, the zoo may continue to be operated until the said application is finally decided or withdrawn and in case of refusal for a further period of six months from the date of such refusal.

(2) Every application for recognition of a zoo shall be made to the Authority in such form and on payment of such fee as may be prescribed.

(3) Every recognition shall specify the conditions, if any, subject to which the applicant shall operate the zoo.

(4) No recognition to a zoo shall be granted unless the Authority, having due regard to the interests of protection and conservation of wild life, and such standards, norms and other matters as may be prescribed, is satisfied that recognition should be granted.

(5) No application for recognition of a zoo shall be rejected unless the applicant has been given a reasonable opportunity of being heard.

(6) The Authority may, for reason to be recorded by it, suspend or cancel any recognition granted under sub-section (4):

Provided that no such suspension or cancellation shall be made except after giving the person operating the zoo a reasonable opportunity of being heard.

(7) An appeal from an order refusing to recognise a zoo under sub-section (5) or an order suspending or canceling a recognition under sub-section (6) shall lie to the Central Government:

(8) An appeal under sub-section (7) shall be preferred within thirty days from the date of communication to the applicant, of the order appealed against: Provided that the Central Government may admit any appeal preferred after the expriy of the period aforesaid if it is satisfied that the appellant had sufficient cause for not preferring the appeal.

38I. Acquisition of animals by a zoo. –

Subject to the other provisions of this Act, no zoo shall acquire or transfer any wild animal specified in Schedule I and Schedule 11 except with the previous permission of the Authority.

38J. Prohibition of teasin& etc., in a zoo. –

No person shall tease, molest, injure or feed any animal or cause disturbance to the animals by noise or otherwise, or litter the grounds in a zoo.

TRADE OR COMMERCE IN WILD ANIMALS, ANIMAL ARTICLES AND TROPHIES

39. Wild Animal, etc. to be Government property. -

(1) Every –

(a) wild animal, other than vermin, which is hunted under Sec. 11 or sec.29 or sub-section (6) of sec 35 or kept or [bred in captivity or hunted[1]] in contravention of any provisions of this Act or any rule or order made thereunder, or found dead, or killed by mistake;

(b) animal article, trophy or uncured trophy or meat derived from any wild animal referred to in Cl.(a) in respect of which any offence against this Act or any rule or order made thereunder has been committed;

(c) ivory imported into India and an article made from such ivory in respect of which any offence against this Act or any rule or order made thereunder has been committed.

(d) vehicle, vessel, weapon, trap or tool that has been used for committing an offence and has been seized under the provision of this Act.

shall be the property of the State Government and, where such animal is hunted in a sanctuary or National Park declared by the Central Government such animal or any article, trophy, uncured trophy or meat derived from such animal or any vehicle, vessel, weapon, trap, or tool used in such hunting, shall be the property of Central Government.

(2) Any person who obtains, by any means, the possession of Government property, shall, within forty-eight hours of obtaining such possession, report it to the nearest police station or authorised officer and shall, if so required, hand over such property to the officer in charge of such police station or such authorised officer, as the case may be.

(3) No person shall, without the previous permission in writing of the Chief Wildlife Warden or the authorised officer.

(a) acquire or keep in his possession, custody, or control, or

(b) transfer to any person, whether by way of gift, sale or otherwise, or

(c) destroy or damage such Government property.

40. Declaration. –

(1) Every person having at the commencement of this Act the control, custody, or possession of any captive animal specified in Sch. I or Part 11 of Sch. 11, or any uncured trophy derived from such animal or salted or dried skin of such animal or the musk of a musk deer or the horn of a rhinoceros, shall, within thirty days from the commencement of this Act, declare to the Chief Wildlife-Warden or the authorised officer the number and description of animal, or article of the foregoing description under his control, custody or possession and the place where such animal or article is kept.

(2) No person shall, after the commencement of this Act, acquire, receive, keep in his control, custody or possession, sell, offer for sale, or otherwise transfer or transport any animal specified in Sch. 1 or Part 11 of Sch. 11, any uncured trophy or meat derived from such animal, or the salted or dried skin of such animal or the musk of a deer or the horn of a rhinoceros, except with the previous permission in writing of the Chief Wildlife Warden or the authorised officer.

(3) Nothing in sub-section (1) or sub-section (2) shall apply to a recognised zoo subject to the provisions of section 381 or to a public museum.

(4) The State Government may, by notification, require any person to declare to the Chief Wildlife Warden or the authorised officer any animal, article or trophy (other than the musk of musk deer or the horn of rhinoceros), or salted or dried skin derived from an animal specified in Sch. I or Part 11 of Sch. H in his control, custody or possession in such form, in such manner, and within such time as may be prescribed.

41. Inquiry and preparation of inventories. -

(1) On receipt of a declaration made under sec.40, the Chief Wildlife Warden or the authorised officer may, after such notice, in such manner and at such time as may be prescribed,

(a) enter upon the premises of a person referred to in Sec.40;

(b) make inquiries and prepare inventories of animal articles, trophies, uncured trophies, salted and dried skins, and captive animals specified in Sch. I and Part 11 of Sch. 11 and found thereon; and

(c) affix upon the animals, animal articles, trophies of uncured trophies, identification marks in such manner as may be prescribed.

(2) No person shall obliterate or counterfeit any identification mark referred to in this Chapter.

42. Certificate of ownership. -

The Chief Wildlife Warden may, for the purposes of Sec. 40, issue a certificate of ownership in such form, as may be prescribed, to any person who, in his opinion, is in lawful possession of any wild animal or any animal article, trophy, or uncured trophy, and may, where possible, mark, in the prescribed manner, such animal article, trophy or uncured trophy for the purposes of Identification.

43. Regulation of transfer of animal etc. -

(1) Subject to the provisions of sub-section (2), (3) and (4), a person (other than a dealer) who does not possess a certificate of ownership shall not

(a) sell or offer for sale or transfer whether by way of sale, gift or otherwise, any wild animal specified in Sch. I or Part II of Sch. 11 or any captive animal belonging to that category or any animal article, trophy, uncured trophy or meat derived therefrom;

(b) make animal articles containing part or whole of such animal;

(c) put under a process of taxidermy an uncured trophy of such animal; except with the previous permission in writing of the Chief Wildlife Warden or the authorised officer.

(2) Where a person transfers or transports from the State in which he resides to another State or acquires by transfer from outside the State any such animal, animal article, trophy or uncured trophy as referred to in sub-section (1) in respect of which he has a certificate of ownership, he shall, within thirty days of the transfer or transport, report the transfer or transport to the Chief Wildlife Warden or the authorised officer within whose jurisdiction the transfer or transport is effected.

(3) No person who does not possess a certificate of ownership shall transfer or transport from one State to another state or acquire by transfer from outside the State any such animal, animal article, trophy or uncured trophy as is referred to in sub-section (1) except with the previous permission in writing of the Chief Wildlife Warden or the authorised officer within whose jurisdiction the transfer is to be effected.

(4) Before granting any permission under sub-section (1) or sub-section (3), the Chief Wildlife Warden or the authorised officer shall satisfy himself that the animal or article referred to therein has been lawfully acquired.

(5) While permitting the transfer or transport or any animal, animal article, trophy or uncured trophy, as is referred to in sub-section (1), the Chief Wildlife Warden or the authorised officer

(a) shall issue a certificate of ownership after such inquiry as he may deem fit;

(b) shall, where the certificate of ownership existed in the name of the previous owner, issue a fresh certificate of ownership in the name of the person to whom the transfer has been effected,.

(c) may affix an identification mark on any such animal, animal article, trophy or uncured trophy.

6. Nothing in this section shall apply –

(a) to tail feather of peacock and animal articles or trophies made therefrom,

(b) to any transaction entered into by a recognised zoo subject to the provisions of Sec 381 or by a public museum with any other recognised zoo or public museum.

44. Dealings in trophy and animal articles without licence prohibited. –

(1) Subject to the provisions of Chapter VA, no person shall, except under, and in accordance with, a licence granted under sub-section (4), –

(a) commence or carry on the business as

(i) a manufacturer of, or dealer in, any animal article, or

(ia) Omitted

(ii) a taxidermist; or

(iii) a dealer in trophy or uncured trophy; or

(iv) a dealer in captive animal; or

(v) a dealer an meat; or

(b) cook or serve meat in any eating-house

(c) derive, collect or prepare, or deal in, snake venom:

Provided that nothing in this sub-section shall prevent a person, who immediately before the commencement of this Act was carrying on the business or occupation specified in this sub-section, from carrying on such business or occupation for a period of thirty days from such commencement, or where he has made an application within that period for the grant of a licence to him, until the licence is granted to him or he is informed in writing that a licence cannot be granted to him.

Provided further that nothing in this sub-section shall apply to the dealers in tail feathers of peacock and articles made therefrom and the manufacturers of such article.

Explanation. – For the purposes of this section, "eating-house" includes a hotel, restaurant or any other place where any eatable is served on payment, whether or not such payment is separately made for such eatable or is included in the amount charged in board and lodging.

(2) Every manufacturer of, or dealer in animal article, or every dealer in captive animals, trophies or uncured trophies, or every taxidermist shall, within fifteen days from the commencement of this Act, declare to the Chief Wildlife Warden his stocks of animal articles, captive animals, trophies and uncured trophies, as the case may be, as on the date of such declaration and the Chief Wildlife Warden or the authorised officer may place an identification mark on every animal article, captive animal trophy or uncured trophy, as the case may be.

(3) Every person referred to in sub-section (1) who intends to obtain a licence shall[11], make an application to the Chief Wildlife Warden or the authorised officer for the grant of a licence.

(4) (a) Every application referred to in sub-section (3) shall be made in such form and on payment of such as fee as may be prescribed, to the Chief Wildlife Warden or the authorised officer.

(b) No licence referred to in sub-section (1) shall be granted unless the Chief Wildlife Warden or the authorised officer having regard to the antecedents and previous experience of the applicant, the implications which the grant of such licence would have on the status of wildlife and to such other matters as may be prescribed in this behalf and after making such inquiry in respect of those matters as he may think fit, is satisfied that the licence should be granted.

(5) Every licence granted under this section shall specify the premises in which and the conditions, if any, subject to which the licensee shall carry on his business.

(6) Every licence granted under this section shall

(a) be valid for one year from the date of its grant;

(b) not be transferable; and

(c) be renewable for a period not exceeding one year at a time.

(7) No application for the renewal of a licence shall be rejected unless the holder of such licence has been given a reasonable opportunity of presenting his case and unless the Chief Wildlife Warden or the authorised officer is satisfied that.

(i) the application for such renewal has been made after the expiry of the period specified thereof, or

(ii) any statement made by the applicant at the time of the grant or renewal of the licence was incorrect of false in material particulars, or

(iii) the applicant has contravened any term of condition of the licence, or any provision of this Act, or any rule made thereunder, or

(iv) the applicant does not fulfill the prescribed conditions.

(8) Every order granting or rejecting in application for the grant or renewal of a licence shall be made in writing.

(9) Nothing in the foregoing sub-section shall apply in relation to vermin.

45. Suspension or calculation of licence. -

Subject to any general or special order of the State Government the Chief Wildlife Warden or the authorised officer may, for reason to be recorded by him in writing, suspend or cancel any licence granted or renewed under sec.44:

Provided that no such suspension or cancellation shall be made except after giving the holder of the licence a reasonable opportunity of being heard.

46. Appeal. –

(1) An appeal from an order refusing to grant or renew a licence under sec.44 of an order suspending or canceling a licence under sec. 45 shall lie

(a) if the order is made by the aurhorised officer, to the Chief Wildlife Warden, or

(b) if the order is made by the Chief Wildlife Warden to the State Government.

(2) In the case of an order passed in appeal by the Chief Wildlife Warden' under C1. (a) of sub-section (1), a second appeal shall lie to the State Government.

(3) Subject as a aforesaid, every order passed in appeal under this section shall be final.

(4) An appeal under this section shall be preferred within days from the date of the communication, to be applicant, or the order appealed against.

Provided that the appellate authority may appeal preferred after the expiry of the period aforesaid if it is satisfied that the appellant had sufficient cause for not preferring the appeal in time.

47. Maintenance of records. -

A licensee under this Chapter shall-

(a) keep records, and submit such returns of his dealings, as may be prescribed.

(i) to the Director or any other officer authorised by him in·this behalf, and

(ii) to the Chief Wildlife Warden or the authorised officer, and

(b) make such records available on demand for inspection by such officers.

48. Purchase of animal, etc., by licensee. - No licensee under this Chapter shall

(a) keep in his control, custody, or possession

(i) any animal, animal article, trophy or uncured trophy in respect of which a declaration under the provisions of sub-section (2) of sec.44 has to be made but has not been made;

(ii) any animal or animal article, trophy, uncured trophy or meat which has not been lawfully acquired under the provisions of this Act or any rule of order made thereunder.

(b) (i) capture any wild animal, or

(ii) acquire, receive, keep in his control, custody, or possession, or sell, offer for sale, or transport, any captive animal specified in Sch. I or Part II of Sch. 11 or any animal article, trophy or uncured trophy, or meat derived therefrom, or serve such meat, or put under a process or taxidening or make animal article containing part or whole of such animal, except in accordance with such rules as may be made under this Act:

Provided that where the acquisition, or possession, or control, or custody of such animal or animal article, trophy or uncured trophy entails the transfer or transport from one State to another, no such transfer or transport shall be effected except with the previous permission in writing of the Director or any other officer authorised by him in this behalf.

Provided further that no such permission under the foregoing proviso shall be granted unless the Director or the officer authorised by him is satisfied that the animal or article aforesaid has been lawfully acquired.

48A Restriction of transportation of wildlife. -

No person shall accept any wild animal (other than vermin) or any animal article, or any specified plant or part or derivative thereof, for transportation except after exercising due care to ascertain that permission from the Chief Wildlife Warden or any other officer authorised by the State Government in this behalf has been obtained for such transportation.

49. Purchase of captive animal, etc. person other than a licensee. -

No person shall purchase, receive or acquire any captive animal, wild animal other than vermin, or any animal article, trophy, uncured trophy, or meat derived therefrom otherwise than from a dealer or from a person authorised to sell or otherwise transfer the same under this Act.

3

Prohibition of Animal Trade

49A. Definitions. -

In this Chapter, –

(a) "scheduled animal" means an animal specified for the time being in Sch. I or Part 11 of Sch. 11;

(b) "scheduled animal article" means an article made from any scheduled animal and includes an article or object in which the whole or any part of such animal (has been used but does not include tail-feather of peacock, an article or trophy made therefrom and snake venom or its derivative;

(c) "specified date" means -

(i) In relation to a scheduled animal on the commencement of the Wildlife (Protection) Amendment Act, 1986, the date of expiry of two months from such commencement.

(ii) in relation to any animal added or transferred to Sch. I or part II of Sch. 11 at any time after such commencement, the date of expiry of two months such addition or transfer;

(iii) in relation to ivory imported into India or an article made from such ivory, the date of expiry of 6 months from the commencement of the Wildlife (Protection Amendment) Act 1991.[3]]

49B. Prohibition of dealing in trophies, animal articles etc. derived from Scheduled animals.-

(1) Subject to the other provisions of this section, on and after the specified date, no person shall

(a) commence or carry on the business as –

(i) a manufacturer of, or dealer, in scheduled animal articles; or

(ia) a dealer in ivory imported into India or article made therefrom or a manufacturer of such article; or

(ii) a taxidermist with respect to any schedule animals or any parts of such animals; or

(iii) a dealer in trophy or unucred trophy derived from any scheduled animal; or

(iv) a dealer in any captive animal being scheduled animal; or

(v) a dealer in meat derived from any scheduled animal[5]; or

(b) cook or serve meat derived from any scheduled animal in any eating-house.

Explanation. - For the purposes of this sub-section, "eating-house" has the same meaning as in the Explanation below sub-section (1) of Sec.44.

(2) Subject to the other provisions of this section, no licence granted or renewed under sec. 44 before the specified date shall entitle the holder thereof or any other person to commence or carry on the business referred to in Cl. (a) of sub-section (1) of this section on the occupation referred into Cl (b) of that sub-section after such date.

(3) Notwithstanding anything contained in sub-section (1) or sub-section (2), where the Central Government is satisfied that it is necessary or expedient to do so in the public interest, it may, by general or special order published in the official Gazette, exempt, for purposes of export, any corporation owned or controlled by the Central Government (including a Government company within the meaning of Sec.617 of the Companies Act, 1956 (1 of 1956), or any society registered under the Societies Registration Act, 1860 (21 of 1860), or any other law for -the time being in force, wholly or substantially financed by the Central Government, from the provisions of sub-section (1) and (2).

(4) Notwithstanding anything contained in sub-section (1) or sub-section (2), but subject to any rules which may be made in this behalf, a person holding a licence under sec-44 to carry on the business as a taxidermist may put under a process of taxidermy any scheduled animal or any part thereof,

(a) for or on behalf of the Government or any corporation or society exempted under sub-section (3), or

(b) with the previous authorisation in writing of the Chief Wildlife Warden, for and on behalf of any person for educational or scientific purposes.

49C. Declaration by dealer. - (1) Every person carrying on the business or occupation referred to in sub-section (1) of Sec.49B shall, within thirty days from the specified date, declare to the Chief Wildlife Warden or the authorised officer,

(a) his stock, if any, as at the end of the specified date of-

(i) scheduled animal articles;

(ii) scheduled animals and part thereof;

(iii) trophies and uncured trophies derived from scheduled animals;

(iv) captive animals, being scheduled animals;

(v) ivory imported into India or article made therefrom.]

(b) the place of places at which the stocks mentioned in the declaration are kept; and

(c) the description of such items, if any, of the stocks mentioned in the declaration which he desires, to retain with himself for his bohafide personal use.

(2) On receipt of a declaration under sub-section (1), the Chief Wildlife Warden or the authorised officer may take all or any of the measures specified in Sec.41 and for this purpose, the provisions of Sec. 41 shall , so far as may be, apply.

(3) Where, in a declaration made under sub-section (1), the person making the declaration expresses his desire to retain with himself any of the items of the stocks specified in the

declaration for his bona fide personal use, the Chief Wildlife Warden, with the prior approval of the Director, may, if he is satisfied that the person is in lawful possession of such items, issue certificates of ownership in favour of such person with respect to all, or as the case may be, such of the items as in the opinion of the Chief Wildlife Warden are required for the bana fide personal use of such person and affix upon such items identification marks in such manner as may be prescribed. Provided that no such item shall be kept in any commercial premises.

(4) No person shall obliterate or counterfeit any identification mark referred to in sub-section (3),

(5) An appeal shall lie against any refusal to grant certificate of ownership under sub-section (3) and the provisions of sub-section (2), (3) and (4) of sec. 46 shall, so far as may be, apply in relation to appeals under this sub-section.

(6) Where a person who has been issued a certificate of ownership under sub-section (3) in respect of any item,

(a) transfers such item to any person, whether by way of gift, sale of otherwise, or

(b) transfer or transports from the State in which he resides to another State any such item, he shall, within thirty days of such transfer or transport, report the transfer or transport to the Chief Wildlife Warden or the authorised officer within whose jurisdiction the transfer or transport is effected.

(7) No person, other than a person who has been issued a certificate of ownership under sub-section (3) shall, on and after the specified date, keep under his control, sell or offer for sale or transfer to [any person any scheduled animal or scheduled animal article or ivory imported into India or any article made therefrom.

PREVENTION AND DETECTION OF OFFENCES

50. Power of entry, search, arrest and detention.

(1) Notwithstanding anything contained in any other law for the time being in force, the Director or any other authorised

by him in this behalf or the Chief Wildlife Warden or the authorised officer or any forest officer or any police officer not below the rank of a sub-inspector may, if he has reasonable grounds for believing that any person has committed an offence against this Act,

(a) require any such person to produce for inspection any captive animal, wild animal, animal article, meat, [[1]trophy, uncured trophy, or any specified plant or part or derivative thereof] in his control, custody or possession, or any licence, permit or any other document granted to him or required to be kept by him under the provisions of this Act;

(b) stop any vehicle or vessel in order to conduct search or inquiry or enter upon and search any premises, land, vehicle, or vessel in the occupation of such person, and open and search any baggage or other things in his possession;

(c) seize any captive animal, wild animal, animal article, meat, trophy or uncured trophy, or any specified plant or part or derivative thereof in respect of which an offence against this Act appears to have been committed, in the possession of any person together with any trap, tool, vehicle, vessel, or weapon used for committing any such offence and unless he is satisfied that such person will appear and answer any charge which may be preferred against him arrest him without warrant and detain him. Provided that where a fisherman, residing within ten kilometres of a sanctuary or National Park , inadvertently enters on a boat not used for commercial fishing, in the territorial waters in that sanctuary or National Park, a fishing tackle or net no such boat shall not be seized.

(2) Omitted 19911

(3) It shall be lawful for any of the officers referred to in sub-section (1) to stop and detain any person, whom he sees doing any act for which a licence or permit is required under the provisions of this Act, for the purposes of requiring such person to produce the licence or permit and if such person fails to produce the licence or permit, as the case may be, he may be arrested without warrant, unless he furnishes his name and address, and otherwise satisfies the officer arresting him that

he will duly answer any summons or other proceedings which may be taken against him.

(3A) Any officer of a rank not inferior to that of an Assistant Director of Wildlife Preservation or Wildlife Warden, who, or whose subordinate, has seized any captive animal or wild animal under clause (c) of sub-section (1) may give the same for custody on the execution by any person of a bond for the production of such animal if and when so required, before the magistrate having jurisdiction to try the offence on account of which the seizure has been made.

(4) Any person detained, or things seized under the foregoing power, shall forthwith be taken before a Magistrate to be dealt with according to law.

(5) Any person who, without reasonable cause, fails to produce anything which he is required to produce under this section, shall be guilty of an offence against the Act.

(6) (a) Where any meat or uncured trophy, specified plant or part or derivative thereof is seized under the provisions of this section, the Asstt. Director of Wildlife Preservation or any other officer of a gazetted rank authorised by him in this behalf or the Chief Wildlife Warden or the authorised officer may arrange for the sale of the same and deal with the proceeds of such sale in such manner as may be prescribed.

(b) Where it is proved that the [[6]meat or uncured trophy, specified plant or part or derivative thereof] seized under the provision of this section is not Government property, the proceed of the sale shall be returned to the owner.

(7) Whenever any person is approached by any of the officers referred to in sub-section (1) for assistance in the prevention or detection of an offence against this Act, or in apprehending persons charged with the violation of this Act, or for seizure in accordance with Ct. (c) of sub-section (1), it shall be the duty of such person or persons to render such assistance.

(8) Notwithstanding anything contained in any other law for the time being in force, any officer not below the rank of an Assistant Director of Wildlife Preservation or Wildlife Warden

shall have the powers, for the purpose of making investigation into any offence against any provision of this Act.

(a) to issue a search warrant;

(b) to enforce the attendance of witiness;

(c) to compel the discovery and production of documents and material objects, and;

(d) to receive and record evidence.

(9) Any evidence recorded under clause (d) of sub-section (8) shall be admissible in any subsequent trial before a Magistrate provided that it has been taken in presence of the accused person.

51. Penalties. -

(1) Any person who 9[contravenes any provisions of this Act [[10]except Chapter VA and section 38J]] or any rule or order made thereunder or who commits a breach of any of the conditions of any licence or permit granted under this Act, shall be guilty of an offence against this Act, and shall, on conviction, be punishable with imprisonment for a term which may extend to [three years] or with fine which may extend to [twenty five thousand rupees] or with both.

Provided that where the offence committed is in relation to any animal specified in Scheduled I or Part 11 of Sch. 11, or meat of any such animal, animal article, trophy, or uncurled trophy derived from such animal or where offence [relates to hunting in, ox, altering the boundaries of] a sanctuary or a National Park, such offence shall be punishable with imprisonment for a term which shall not be less than [one year] but may extend to six years and also with fine which shall not be less than [five thousand rupees.

[Provided further that in the case of a second or subsequent offence of the nature mentioned in this sub-section, the term or imprisonment may extend to six years ' and shall not be less than two years and the amount of fine shall not be less than ten thousand rupees.]

(1A) Any person who contravenes any provisions of Chapter VA, shall be punishable with imprisonment for a7 term which shall not be less then one year but which may extend to seven years and also with fine which shall not be less than five thousand rupees.]

(1B) Any person who contravenes the provisions of Section 38J] shall be punishable with imprisonment for a term which may extend to six months or with fine which may extend to two thousand rupees, or with both.

Provided that in case of second or subsequent offence the term of imprisonment may extend to one year or the fine may extend to five thousand rupees.

(2) When any person is convicted of an offence against this Act, the Court trying the offence may order that any captive animal, wild animal, animal article, trophy, uncured trophy, meat, ivory imported into India or an article made from such ivory, any specified plant or part or derivative thereof in respect of which the offence has been committed, any trap, tool, vehicle, vessel, or weapon used in the commission of the said offence be forfeited to the State Government and that any licence or permit, held by such person under the provisions of this Act, be cancelled.

(3) Such cancellation of licence or permit o r such forfeiture shall be in addition to any other punishment that may be awarded for such offence.

(4) Where any person is convicted of an offence against this Act, the court may direct that the licence, if any, granted to such person under the Arms Act, 1959 (54 of 1959) for possession of any arm with which an offence against this Act has been committed, shall be cancelled, and that such person shall not be eligible for a licence under the Arms Act, 1959, for a period of five years from the date of conviction.

(5) Nothing contained in section 360 of the Code of Criminal Procedure, 1973 (2 of 1974) or in the Probation of offenders Act, 1958 (20 of 1958) shall apply to a person convicted of an offence with respect to hunting in a sanctuary or a National Park or

of an offence against any provision of Chapter VA unless such person is under eighteen years of age.

52. Attempts and abetment.

Whoever attempts to contravene, or abets the contravention of, any of the provisions of this Act or of any rule of order made thereunder shall be deemed to have contravened that provision or rule or order, as the case may be.

53. Punishment for wrongful seizure.

If any person, exercising powers under this Act, vexatiously and unnecessarily seizes the property of any other person on the pretence of seizing it for the reasons mentioned in sec. 50, he shall, on conviction, be punishable with imprisonment for a term which may extend to six months, or with fine which may extend to five hundred rupees, or with both.

54. Power to compound offences.

(1) The Central Government may; by notification empower the Director of Wildlife Preservation or any other officer and the State Government], may, by notification, empower the C-1-def Wildlife Warden or any officer or a rank not inferior to that of a Deputy Conservator of Forests,

(a) to accept, from any person against whom a reasonable suspicion exists that he has committed an offence against this Act, payment of a sum of money by way of composition of the offence which such person is suspected to have committed; and.

(b) when any property that has been seized is liable to be forfeited, to release the same on payment of the value thereof as estimated by such officer.

(2) On payment of such sum of money or such value, or both, as the case may be, to such officer, the suspected person, if in custody, shall be discharged, and the property, other than Government property, if any seized, shall be released and no further proceedings in respect of the offence shall be taken against such person.

(3) The officer compounding any offence may order the cancellation of any licence or permit granted under this Act to

the offender, or if not empowered to do so, may approach an officer so empowered, for the cancellation of such licence or permit.

(4) The sum of money accepted or agreed to be accepted as composition under C1. (b) of sub-section (1) shall, in no case, exceed the sum of two thousand rupees.

Provided that no offence, for which a minimum period of imprisonment has been prescribed in sub-section (1) of sec 51, shall be compounded.

55 Cognizance of offences.

No court shall take cognizance of any offence against this Act on the complaint of any person other than

(a) the Director of Wildlife Preservation or any other officer authorised in this behalf by the State Government; or

(b) the Chief Wildlife Warden, or any other officer authorised in this behalf by the State Government; or

(c) any person who has given notice of not less than sixty days, in the mariner prescribed, of the alleged offence and of his intention to make a complaint, to the Central Government or the officer authorised as aforesaid.

56. Operation of other laws not barred.

Nothing in this Act shall be deemed to prevent any person from being prosecuted under any other law for the time being in force, for any act or omission which constitutes an offence against this Act or from being liable under such other law to any higher punishment or penalty than that provided by this Act.

Provided that no person shall be punished twice for the same offence.

57. Presumption to be made in certain cases.

Where, in any prosecution for one offence against this Act, it is established that a person is in possession, custody or control of any captive animal, animal article, [meat, trophy, uncured, trophy, specified plant, or part or derivative thereof, it shall be presumed until the contrary is proved, the burden of proving which shall lie on the accused, that such person is in unlawful possession, custody or control of such captive animal, animal article, meat, [trophy, uncured trophy] specified plant, or part or derivative thereof.

58. Offenses by companies. -

(1) Where an offence against this Act has been committed by a company, every person who, at the time the offence was committed, was in charge of, and was responsible to, the company for the conduct of the business of the company as well as the company, shall be deemed to be guilty of the offence and shall be liable to be proceeded against and punished accordingly.

Provided that nothing contained in this sub-section shall. render any such person liable to any punishment, if he proves that the offence was committed without his knowledge or that he exercised all due diligence to prevent the commission of such offence.

(2) Notwithstanding anything contained in sub-section (1), where an offence against this Act has been committed by a company and it is proved that the offence has been committed with the consent or connivance of, or is attributable to any neglect on the part of any director, manager, secretary, or other officer shall also be deemed to be guilty of that offence and shall be liable to be proceeded against and punished accordingly.

Explanation. - For the purpose of this Section,

(a) "company' means any body corporate and includes a firm or other association of individuals; and

(b) "director', in relation to. a firm, means a partner in the firm

MISCELLANEOUS

59. Officers to be public servants.

Every officer referred to [in Chapter II and the chairperson, members, member-secretary, officers and other employees referred to in chapter IVA[1]] and every other officer exercising any of the powers conferred by this Act shall be deemed to be a public servant within the meaning of sec.21 of the Indian Pedal Code (45 of 1860).

60. Protection of action taken in good faith. -

(1) No suit, prosecution, or other legal proceeding shall lie against any officer or other employee of the Central Government or the State Government for anything which is in good faith done or intended to be done under this Act.

(2) No suit or other legal proceeding shall lie against the Central Government or the State Government or any of its officers or other employees, for any damage caused or likely to be caused by anything which is in good faith done or intended to be done under this Act.

(3) No suit or other legal proceeding shall lie against the Authority referred to in Chapter IVA and its chairperson, member, member secretary, officers and other employees for any thing which is in good faith done or intended to be done under this Act.]

(60A) Reward to persons. -

(1) When a court imposes a sentence of fine or a sentence of which fine forms a part, the court may when passing judgement order that the reward be paid to a person who renders assistance in the detection of the offence or the apprehension of the offenders out of the proceeds of fine not exceeding twenty percent of such fine.

(2) When a cage is compounded under section 54, the officer compounding may order reward to be paid to a person who renders assistance in the detection of the offence or the apprehension of the offenders out of the sum of money accepted by way of composition not exceeding twenty percent of such money].

61. Power to alter entries in schedules. –

(41) The Central Government may, if it is of the opinion that it is expedient so to do, by notification, [[4]add or delete any entry to or from any schedule] or transfer any entry from one

part of the schedule to another part of the same schedule or from one schedule to another.

(2) Omitted 1991

(3) On the issue of a notification under sub-section (1) the relevant schedule shall be deemed to be altered accordingly, provided that every such alteration shall be without prejudice to anything done or omitted to be done before such alteration.

(4) Omitted 1991

62. Declaration of certain wild animal to be vermin. -

The Central Government] may by notification, declare any wild animal other than those specified in Sch. I and part 11 of Sch H to be vermin for any area and for such period as may be specified therein and so long as such notification is in force, such wild animal shall be deemed to have been included in Sch.V.

63. Power of Central Government to make rules. -

(1) The Central Government may, by notification, make rules for all or any of the following matters, namely:

(a) conditions and other matters subject to which a licensee may keep any specified plant in his custody or possession under section 17F;

(b) the salaries and allowances and other conditions of appointment of chairperson, members and members-secretary under sub-section (5) of Section 3813;

(c) the terms and conditions of service of the officer and other employees of the Central Zoo Authority under sub-section (7) of section 3813;

(d) the form in which the annual statement of accounts of Central Zoo Authority shall be prepared under sub-section (4) of Section 38E;

(e) the form in which and the time at which the annual report of Central Zoo Authority shall be prepared under section 38F;

(f) the form in which and the fee required to be paid with application for recognition of a zoo under sub-section (2) of section 38H;

(g) the standards, norms and other matters to be considered for granting recognition under sub-section (4) of section 38H;

(h) the form in which declaration shall be made under sub-section (2) of section 44;

(i) the matters to be prescribed under clause (b) sub-section (4) of section 44;

(j) the terms and conditions which shall govern transaction referred to in clause (b) of section 48;

(k) the manner in which notice may be given by a person under clause (c) of section 55;

(1) the matters specified in sub-section (2) of section 64 in so far as they relate to sanctuaries and National Parks declared by the Central Government]

(2) Every rule made under this section shall be laid, as soon as may be, after it is made, before each House of Parliament, while it is in session, for a total period of thirty days which may be comprised in one session or in two or more successive sessions and if before the expiry of the session immediately following the session or the successive session aforesaid, both Houses agree in making any modification in the rule or both Houses agree that the rule should not be made, the rule shall thereafter have effect only in such modified form or be of no effect, as the case may be; so however, that any such modification or annulment shall be without prejudice to the validity of anything previously done under that rule.

64. Power of State Government to make rules.– (1) The State Government may, by notification, make rules for carrying out the provisions of this Act in respect of matters which do not fall within the purview of Sec.63

(2) In particular and without prejudice to the generality of the foregoing power, such rules may provide for all or any of the following matters, namely:

(a) the term of office of the members of the Board referred to in Cl. (g) of sub-section (1) of Sec.6 and the manner of filling vacancies among them;

(b) allowances referred to in sub-section (4) of Sec.6;

(c) the forms to be used for any application, certificate, claim, declaration, licence, permit, registration, return, or other document, made, granted, or submitted under the provisions of this Act and the fees, if, any therefor;

(d) the conditions subject to which any licence or permit may be granted under this Act;

(e) the particulars of the record of wild animal (captured or killed) to be kept and submitted by the licensee;

10(ee) the manner in which measures for immunization of live-stock shall be taken;

(f) regulation of the possession, transfer, and the sale of captive animals, meat, animal articles, trophies, and uncured trophies;

(g) regulation of taxidermy;

(h) any other matter which has to be, or may be, prescribed under this Act.

65. Rights of Scheduled Tribes to be protected. -

Nothing in this Act shall affect the hunting rights conferred on the Scheduled Tribes of the Nicobar Islands in the Union Territory of Andaman and Nicobar Islands by notification of the Andaman and Nicobar Administration, NO. 40/97/1'. No.G-635. Vol. III, dated the 28th April, 1967 published at pages I to 5 Extraordinary issue of the Andaman and Nicobar Gazette, dated 28th April, 1967.

66. Repeal and savings. –

(1) As from the commencement of this Act, every other Act relating to any matter contained in this Act and in force in a State shall, to the extent to which that Act or -any provision contained therein corresponds, or is repugnant, to this Act or any provision contained in this Act, stand repealed :

Provided that such repeal shall not–

(i) affect previous' operation of the Act so repealed, or any thing duly done or suffered thereunder;

(ii) affect any right, privilege, obligation, or liability acquired, accrued, or incurred under the Act so repealed;

(iii) affect any penalty, forfeiture or punishment incurred in respect of any offence committed against the Act so repealed; or

(iv) affect any investigation, legal proceeding, or remedy in respect of any such right, privilege, obligation, liability, penalty, forfeiture, or punishment as aforesaid; and any such investigation, legal proceeding, or remedy may be instituted, continued, or enforced, and any such penalty, forfeiture, and punishment may be imposed, as if the aforesaid Act had not been repealed.

(2) Notwithstanding such repeal,

(a) anything don0l or nay action taken under the Act so repealed (including any notification, order, certificate, notice, or receipt issued, application made, or permit granted) which is not inconsistent with the provisions of this Act be deemed to have been done or taken under the corresponding provisions of this Act as if this Act were in force at the time such thing was done or action was taken, and shall continue to be in force, unless and until superseded by anything done or an action taken under this Act;

(b) every licence granted under any Act so repealed and in force immediately before the commencement of this Act shall be deemed to have been granted under the corresponding provisions of this Act and shall, subject to the provisions of this Act, continue to be in force for the unexpired portion of the period for which such licence had been granted.

(3) For the removal of doubts, it is hereby declared that any sanctuary or National Park declared by a State Government under any Act repealed under sub-section (1) shall be deemed to be a sanctuary or National Park, as the case may be, declared by the State Government under this Act and where any right

in or over any land in any such National Park which has not been extinguished under the said Act, at or before the commencement of this Act, the extinguishment of such rights, shall be made in accordance with the provisions of this Act.

(4) For the removal of doubts, it is hereby further declared that where any proceeding under any provision of Sections 19 to 25 (both inclusive) is pending on the date of commencement of the Wildlife (Protection) Amendment Act, 1991, any reserved forest or a part of territorial waters comprised within a sanctuary declared under section 18 to be a sanctuary before the date of such commencement shall be deemed to be a sanctuary, before the date of such commencement shall be deemed to be a sanctuary declared under section 26A.

S.O.1092 (E) – In exercise of the powers conferred by sub-section (2) and (3) of section 5A read with section 63 of the Wild Life (Protection) Act, 1972 (53 of 1972), the Central Government hereby makes the following rules, namely:-

1. Short title and commencement –

(1) These rules may be called the National Board for Wild Life Rules, 2003

(2) They shall come into force on the date of their publication in the Official Gazette.

2. Definitions –

In these rules, unless the context otherwise requires:-

(a) "Act" means the Wild Life (Protection) Act, 1972 (53 of 1972);

(b) "National Board" means the National Board for Wild Life constituted under section 5A of the Act;

(c) "Chairperson" means the Chairperson of the National Board;

(d) All other words and expressions used in these rules shall have the meanings respectively assigned to them in the Act.

3. National Board for Wild Life -

(1) The National Board shall be constituted in accordance with the provisions contained in sub-section (1) of section 5A of the Act.

(2) The term of office of the members other than those who are members ex-officio shall be for a period of three years.

(3) The Member of Parliament nominated to the National Board shall cease to be the Member of the National Board on his having ceased to the Member of the Parliament.

(4) The vacancies referred to in clauses (e), (f) and (v) of sub-section (1) of section 5A of the Act shall be filled by nomination by the Central Government with the approval of chairperson.

4. Meetings -

(1) The National Board shall ordinarily meet once in a year at a place to be decided with the approval of the Chairperson.

(2) The Member-Secretary shall prepare agenda items for the meetings, obtain approval of the Chairperson and circulate it to all members at least fifteen days prior to the date of such meeting.

(3) The Member-Secretary shall prepare the minutes of the meeting and circulate it to all members within thirty days of the meeting after obtaining approval of the Chairperson.

(4) The Member-Secretary shall also initiate follow up action on the decisions of the National Board.

5. Removal from the National Board -

The Central Government may, with the prior approval of the chairperson, remove any member of the National Board appointed under clause (e), (f) and (v) of sub-section (1) of section 5A of the Act, if-

(a) he is of unsound mind and stands to declared by a competent court; or

(b) he is an undischarged insolvent, or

(c) he is convicted of any offence including moral turpitude; or

(d) his conduct is found to be unbecoming of a member.

6. Disclosure of interest by members -

If any member of the National Board or his family member has any interest in a proposal submitted for consideration of the National Board, the member shall disclose the extent of his interest in the proposal.

7. Presiding over the meetings of the National Board -

(1) The Chairperson or in his absence the Vice Chairperson shall preside at every meeting of the National Board.

(2) In the absence of both the Chairperson and the Vice-Chairperson, such other member present as is decided by the National Board shall preside over the meeting.

8. Quorum –

(1) The quorum necessary for the transaction of business at a meeting of the National Board shall be one-half of the total number of members.

(2) If at any time there is no quorum, the Chairperson or the person presiding over a meeting shall adjourn the meeting until there is a quorum.

9. Invitees to the National Board meetings -

The chairperson of the National Board may invite any person or persons having experience in the wild life conservation to attend any meeting of the National Board.

10. Record of Business -

(1) The record of business transaction at every meeting of the National Board shall be prepared by the Member-Secretary for approval by the Chairperson or the Member presiding at such meeting.

(2) When any business is transacted by circulation of papers, the Member-Secretary shall prepare a record of the business so transacted for approval by the Chairperson of the National Board permitting the circulation.

(3) The record of business transacted at every meeting of the National Board shall be approved and signed by the Chairperson or the Vice-Chairperson or the member presiding

at such meeting, and the approved record of business shall be circulated to all the members.

(4) All records shall be maintained by the Member-Secretary of items of business transacted by the National Board or the committees, sub-committees or sub-groups thereof.

11. Appointment of Committees -

(1) The National Board may, constitute a Standing Committee and such other committees or study groups, as may be necessary, in accordance with the provisions contained in section 5B of the Act.

(2) The National Board shall determine the terms of reference, tenure, membership and other related aspects as may be necessary in rspect of the committees constituted under sub-rule (1) of this rule.

12. Traveling and other allowances to members of the National Board and its committees -

(1) A member of the National Board or any committee other than member ex-officio, shall be entitled to receive, in respect of any journey performed by him for the purpose of attending a meeting of the National Board or of a duly constituted committee thereof or for the purpose of discharging any duty assigned to him by the National Board or the committee concerned, traveling allowances and daily allowances at the rates as are admissible to the Group 'A' officers of the central Government for the time being in force.

(2) No traveling allowance or daily allowance shall be allowed to a member of the National Board or of any committee unless he certifies that he has not drawn any traveling or daily allowance from any other source in respect of the journey and halt for which the claim is made.

(3) The traveling allowance shall be payable from the usual place of residence of a member of the National Board or any of the committee to the place of the meeting or the place where he has gone to attend to any business of the National Board and back to his place of residence.

(4) In special circumstances, the chairperson may grant traveling allowance from the places other than usual place of residence of a member.

13. Interpretation of rules -

Where any doubt arises as to the interpretation of these rules, interpretation shall be referred to the Central Government for its decision.

S.O. 445 (E) – In exercise of the powers conferred by sub-sections (1) and (3) of section 40A read with section 63 of the Wild Life (Protection) Act, 1972 (53 of 1972), the Central Government hereby makes the following rules, namely:

1. Short title and commencement –

(1) These rules may be called the declaration of Wild Life Stock Rules, 2003.

(2) They shall come into force on the date of their publication in the Official Gazette.

2. Definitions:

In these rules, unless the context otherwise requires, -

(a) "Act" means the Wildlife (Protection) Act, 1972 (52 of 1972);

(b) "form" means the form annexed to these rules;

(c) all other words and expressions used in these rules shall have the meanings respectively assigned to them in the Act.

3. Publicity of intent of notification and Assistance in making application:-

(1) The Chief Wildlife Warden or the officer authorized by the State Government in this regard shall cause to give wide publicity to the intent of this notification in the regional language through electronic or print media or such other means.

(2) The Chief Wildlife Warden or the officer authorized by the State Government in this regard shall take necessary action to assist the local communities and individuals especially the poor and illiterate in the declaration of their possession, filling

up the specified form and any other matter connected therewith and shall make every attempt to ensure that no individual or community associated with animals is deprived of this opportunity.

4. Procedure for filing applications –

(1) An application to the Chief Wildlife Warden or the officer authorized by the State Government in this regard shall be presented in the Form annexed to these rules by the applicant either in person or by an agent or by duly authorized legal practitioner or sent by registered post addressed to the Chief Wild Life Warden or the officer authorized by the Sate Government in this regard of the concerned State or the Union territory.

(2) The application under sub rule (1) shall be presented in four complete sets within a period of one hundred and eighty day from the date of publication of these rules.

(3) The applicant may attach to and present with his application an acknowledgement slip as is given in the form which shall be signed by the official receiving the application on behalf of the Chief Wild Life Warden or the officer authorized by the State Government in this regard in acknowledgement of the receipt of the application.

5 Presentation and scrutiny of applications :-

(1) The Chief Wildlife Warden or the officer authorized by the State Government in this regard shall endorse on every application the date on which it is presented or deemed to have been presented under that rule and shall sign the endorsement.

(2) If on scrutiny, the application is found to be in order, it shall be duly registered and give serial number.

(3) If the application, on scrutiny, is found to be defective, the same shall be returned to the applicant within fifteen days for rectifying the defects and resubmitting the corrected application within fifteen days from the date of its receipt.

(4) If the applicant fails to rectify the defect within the time allowed under sub-rule (3), the Chief Wild Life Warden or the

officer authorized by the State Government in this regard may, by order and for the reasons to be recorded in writing, decline to register the application.

6 Place of filing application :-

The applicant shall file application with the Chief Wild Life Warden or the officer authorized by the State Government in this regard.

7 Date and Place of hearing to be notified –

The Chief Wild Life Warden or the officer authorized by the State Government in this regard shall notify to the parties the date, place and time of hearing of each application, if required.

8 Decision on applications –

(1) The Chief Wild Life Warden or the officer authorized by the State Government in this regard shall verify the facts mentioned in the application and make such inquiry as maybe required.

(2) The Chief Wild Life Warden shall, as far as possible, decide the application within six months of the date of its presentation and communicate the same to the applicant in writing under his own signature by registered post.

9 Hearing on application ex-parte –

Where on the date fixed for hearing the application, the applicant fails to appear without intimation, the Chief Wild Life Warden or the officer authorized by the State Government in this regard may at their discretion adjourn or decide the application ex-parte.

10 Inquiry by the Chief Wild Life Warden or Authorized Officer –

(1) The Chief Wild Life Warden or the officer authorized by the State Government in this regard shall conduct a detailed inquiry and take all actions as provided in section 41 of the Act.

(2) A copy of the report pertaining to sub-rule (1) f this rule, shall be provided to the applicant.

11. Certificate of ownership –

(1) The Chief Wild Life Warden shall provide a certificate of ownership to the applicant whose claim is found valid.

(2) The certificate of ownership shall be provided as per the provisions of section 42 of the Act.

(3) The certificate of ownership shall contain the facsimile of the identification mark and in case of live animals the identification number of the transponder (microchip) implanted shall be mentioned in the certificate.

12. Dealing with declared objects:

Any captive animal, animal article, trophy or uncured trophy declared under sub section (1) of section 40A and in respect of which certificate of ownership has not been granted or obtained, shall be treated as government property.

13. Order to be signed and dated –

Every order of the Chief Wild Life Warden shall be in writing and shall be signed and dated by the Chief Wild Life Warden.

14. Communication of order to parties:-

Every order passed on the application shall be communicated to the applicant either in person or by registered post free of cost.

G.S.R. 350(E). - In exercise of the powers conferred by Clause (h) of Section 63 read with Section 17E of the Wild Life (Protection) Act, 1972 (53 of 1972), the Central Government hereby makes the following rules, namely:

1. Short title and commencement.- (1) These rules may be called the Wildlife (Specified Plant Stock Declaration) Central Rules, 1995.

(2) These rules shall come into force from the date of commencement of provisions of Chapter IIIA of the Wild Life (Protection) Act, 1972.

2. Declaration, of stocks by a cultivator or dealer in specified plants, parts and derivatives thereof,

Every cultivator of specified plants and the dealer in specified plants and derivatives thereof shall, within 30 days from the commencement of provisions of Chapter IIJA of Wild Life (Protection) Act, 1972, declare his stocks of specified plants, parts and derivatives thereof, as the case may be, as on the date of such declaration to the Chief Wildlife Warden in the form given below:

FORM OF DECLARATION

(See Section 17E and Sub-section 2 of Section 44)

To

The Chief Wildlife Warden,

State of /Union Territory of ....

1 Full name and address of the cultivator or dealer in specified plants, parts and derivatives thereof making the declaration..

2. Actual stock held on the date of declaration:

Name of the specified plant (including scientific name)	Known Uses	Description of stock	Quantity held in stock Kgs number	Premises where stock are kept	Date of procure-ment	Source & specific Area of procurement	Docum entary proof, if any
1	2	3	4	5	6	7	8

3 Remarks, if any ....

I do hereby declare that the information given above is true to the best of my knowledge and belief

Place : Signature of the person

Date : making the declaration

4

Wildlife Specified Plants

G.S.R. 349. (E). - In exercise of powers /conferred by Clause (a) of Subsection (1) of Section 63 of Wild Life (Protection) Act, 1972 (53 of 1972), the Central Government hereby makes the following rules, namely:

1. Short title, Extent and Commencement.-

(1) These Rules may be called Wildlife (Specified Plants - Conditions for possession by licensee) Rules, 1995.

(2) These Rules shall come into force from the date of commencement of provisions of Chapter IIIA of the Wild Life (Protection) Act, 1972.

2. Definition-

In these Rules, unless the context otherwise requires, "Act" means the Wild Life (Protection) Act, 1972 (53 of 1972).

3. Conditions and other matters subject to which the licensee may keep any specified plants in his castody or possession. (1) No licensee shall acquire or receive or keep in his control, custody or possession any specified plant or part or derivative thereof in respect of which a declaration under Section 17E of Act has not been made.

(2) No licensee shall acquire, purchase or receive any specified plant or part or derivative thereof from any person other than a licensed dealer in specified plants or a cultivator having a license for cultivation of specified plants under the Act.

(3) Licensee shall keep the stock of specified plants so purchased by him only in the premises approved by the Chief Wildlife Warden of the State

WILDLIFE (PROTECTION) RULES, 1995

G.S.R. 348(E)-In exercise of powers conferred by clause (k) of sub-section (1) of section 63 of the Wildlife (Protection) Act, 1972 (53 of 1972), the Central Government hereby makes the following rules, namely:

1. Short title and Commencement–

(1) These Rules may be called Wildlife (Protection) Rules, 1995.

(2) They shall come into force from the date of their publication in the Official Gazette.

2. In these rules, unless the context otherwise requires

(a) "Act" means the Wild Life (Protection) Act, 1972.

(b) "Section" means the Section of the Act.

3. The manner of the notice under clause (c) of section 55- (1) The notice to the Central Government or the State Government or any authorised officer, as the case may be, shall be given in Form "A" annexed to these rules.

(2) The person/giving notice to the Central Government or the State Government or any authorised officer shall send the notice by registered post to: (a) The Director of Wildlife Preservation, Government of India in the Ministry of Environment and Forests, New Delhi; and (b) (i) The Secretary to the concerned State Govt./Union Territory incharge of Wildlife, or (ii) The Chief Wildlife Warden of the concerned State Govt./Union Territory, or (iii) Any authorised officer of State Government/Union Teritory.

RECOGNITION OF ZOO RULES 1992

Notification

In exercise of the powers conferred by clauses (f) and (g) of sub-section (i) of Section 63 of the Wild Life (Protection) Act

1972 (53 of 1972), the Central Government hereby makes the following rules, namely:

1. Short title and commencement:

(1) These rules may be called the Recognition of Zoo Rules, 1992.

(2) They shall come into force on the date of their publication in the Official Gazette.

2. Definitions : In these rules, unless the context otherwise requires,

(a) "Act" means the Wild Life (Protection) Act, 1972 (53 of 1972);

(b) "Enclosure" means any accommodation provided for Zoo animals;

(c) "Enclosure barrier" means a physical barrier to contain an animal within an encloure;

(d) "Endangered species" means species included in Schedule I of the Act;

(e) "Form" means form set forth in Appendix A to these rules;

(f) "Performing purposes" means any effort to force the animal to carry out unnatural act including performance of circus tricks;

(g) "Stand-of barrier" means a physical barrier set back from the outer edge of an enclosure barrier;

(h) "Zoo operator" means the person who has ultimate control over the affairs of the Zoo provided that

(i) in the case of firm or other association of individuals, any one of the individual partners or members thereof shall be deemed to be the Zoo operator;

(ii) in the case of a company, any director, manager, secretary or other officer, who is in-charge of and responsible to the company for the affairs of the Zoo shall be deemed to be the Zoo operator;

(iii) in the case of zoo owned or controlled by the Central Government or any State Government, or any local authority, the person or person appointed to manage the affairs of the zoo by the Central Government, the State Government or the local authority, as the case may be shall be deemed to be the Zoo operator.

3. Application for recognition: An application under section 38H of the Act for recognition of a zoo shall be made to the Central Zoo Authority in Form A.

4. Fees for application:

(a) There shall be paid in respect of every application under rule 3 a fee of rupees five hundred.

(b) The amount of the fee shall be paid through Demand Draft/Postal Order(s) in favour of the Central Zoo Authority, New Delhi.

5. Documents to be filed alongwith the application and particulars it should contain: Every application shall be accompanied by the prescribed fee and shall contain clear particulars as to the matters specified in Form A.

6. Power to make inquiries and call for information : Before granting recognition to a zoo under section 38H of the Act, the Central Zoo Authority may make such inquiries and require such further information to be furnished, as it deems necessary, relating to the information furnished by the zoo in its application in Form A.

7. Form of recognition: The recognition granted to a zoo shall be subject to the following conditions, namely:

(a) that the recognition unless granted on a permanent basis, shall be for such period not less than one year as may be specified in the recognition;

(b) that the zoo shall comply with such standards and norms as are or may be prescribed or imposed under the provisions of the Act and these rules from time to time.

8. Renewal of recognition:

(a) Three months before the expiry of the period of recogntion, a recognised zoo desirous of renewal of such recognition may make an application to the Central Zoo Authority in Form A.

(b) The provisions of rules 3, rule 4, rule 5, rule 6 and rule 7 shall apply in relation to renewal of recognition as they apply in relation to grant of recognition except that, the fee payable in respect of an application for renewal of recognition shall be rupees two hundred.

9. Classification of Zoos:

For the purposes of deciding standards and norms for recognition of Zoos and monitoring and evaluating their performance, the zoos, on the basis of the area, number of animals and their variety exhibited, and the number of visitors, shall be classfied into four categories as specified below:

Category of the Zoo	Large	Medium	Small	Mini
Area of the zoo in hectares	More than 75 hectares	50-75 hectare	20-50 hectare	Less than 20 hectare
Number of animals exhibited	More than 750	500-750	200-499	200
Animals variety exhibited	More than 75 numbers	50-75 numbers	20-49 numbers	20 numbers
Number of endangered species exhibited	More than 15	15-Oct	9-May	Less than 5
Annual attendance of visitors per year	More than 7.5 lakhs	5-7.5 lakhs	5-Feb lakhs	Less than 2 lakhs

10. Standards and norms subject to which recognition under section 38H of the Act shall be granted:

The Central Zoo Authority shall grant recognition with due regard to the interests of protection and conservation of wild life, and such standards, norms and other matters as are specified below :

General :

(1) The primary objective of operating any zoo shall be the conservation of wildlife and no zoo shall take up any activity that is inconsistent with the objective.

(2) No zoo shall acquire any animal in violation of the Act or rules made thereunder.

(3) No zoo shall allow any animal to be subjected to the cruelties as defined under the Prevention of Cruelty to Animals Act, 1960 (59 of 1960) or permit any activity that. exposes the animals to unnecessary pain, stress or provocation, including use of animals for performing purposes.

(4) No zoo shall use any animal, other than the elephant in plains and yak in hilly areas for riding purposes or draughting any vehicle.

(5) No zoo shall keep any animal chained or tethered unless doing so is essential for its own well being.

(6) No zoo shall exhibit any animal that is seriously sick, injured or infirm.

(7) Each zoo shall be closed to visitors at least once a week.

(8) Each zoo shall be encompassed by a perimeter wall at least two metres high from the ground level. The existing zoos in the nature of safaries and deer parks will continue to have chain link fence of appropriate design and dimensions.

(9) The zoo operators shall provide a clean and healthy environment in the zoo by planting trees, creating green belts and providing lawns and flower beds etc.

(10) The built up area in any zoo shall not exceed twenty five per cent of the total area of the zoo. The built up area includes administrative buildings, stores, hospitals, restaurants, kiosks and visitor rest sheds etc., animal houses and 'pucca' roads.

(11) No-zoo shall have the residential complexes for the staff within the main campus of the zoo. Such complex, if any, shall be separated from the main campus of the zoo by a boundary wall with a minimum height of two metres from the ground level.

Administrative and Staffing Pattern:

(12) Every zoo shall have one full-time officer in-charge of the zoo. The said officer shall be delegated adequate

administrative and financial powers as may be necessary for proper upkeep and care of zoo animals.

(13) Every large and medium zoo shall have at least one full-time curator having the sole responsibility of looking after the upkeep of animals and maintenance of animal enclosures.

(14) Each large zoo shall have at least two full-time veterinarians and medium and small zoo shall have at least one full-time veterinarian. The mini zoo may at least have arrangement with any outside veternarian for visiting the zoo every day to look after the animals.

Animal Enlosures - Design, Dimensions and other Essential Features:

(15) All animal enclosures in a zoo shall be so designed as to fully ensure the safety of animals, caretakers and the visitors. Stand of barriers and adequate warning signs shall be provided for keeping the visitors at a safe distance from the animals.

(16) All animal enclosures in a zoo shall be so designed as to meet the full biological requirements of the animals housed therein. The enclosures shall be of such size as to ensure that the animals get space for their free movement and exercise and the animals within herds and groups are not unduly dominated by individuals. The zoo operators shall take adequate safeguards to. avoid the animals being unnaturally provoked for the benefit of viewing by public and excessive stress being caused by visibility of the animals in the adjoining enclosures.

(17) The zoo operators shall endeavour to simulate the conditions of the natural habitat of the animal in the enclosures as closely as possible. Planting of appropriate species of trees for providing shade and constructing shelters which would merge in the overall environment of the enclosures, shall also be provided. Wherever it is technically feasible, only moats shall be provided as enclosure barriers.

(18) The enclosures housing the endangered mammalian species, mentioned in Appendix I to these rules, shall have feeding and retiring cubicles/cell of minimum dimensions given in the said appendix. Each cubicle/cell shall have resting, feeding

drinking water and exercising facilities, according to the biological needs of the species. Proper ventilation and lighting for the comfort and well being of animals shall be provided in each cell /cubicle /enclosure.

(19) Proper arrangement of drainage of excess of water and arrangements for removal of excreta and residual water from each cell /cubicle /enclosure shall be made.

(20) Designing of any new enclosure for endangered species shall be finalized in consultation with the Central Zoo Authority.

Hygiene, Feeding and Upkeep:

(21) Every zoo shall ensure timely supply of wholesome and unadulterated food in sufficient quantity to each animal according to the requirement of the individual animals, so that no animal remains undernourished.

(22) Every Zoo shall provide for a proper waste disposal system for treating both the solid and liquid wastes generated in the zoos.

(23) All left over food items, animal excreta and rubbish shall be removed from each enclosure regularly and disposed of in a manner congenial to the general cleanliness of the zoo.

(24) The zoo operators shall make available round the clock supply of potable water for drinking purposes in each cell / enclosure /cubicle.

(25) Periodic application of disinfectants in each enclosure shall be made according to the directions of the authorised veterinary officer of the zoo.

Animal Care, health and Treatment:

(26) The animals shall be handled only by the staff having experience and training in handling the individual animals. Every care shall be taken to avoid discomfort, behavorial stress or physical harm to any animal.

(27) The condition and health of all animals in the zoo shall be checked every day by the person in-charge of their care. if any animal is found sick, injured, or unduly stressed the matter

shall be reported to the veterinary officer for providing treatment expeditiously.

(28) Routine examination including parasite checks shall be carried out regularly and preventive medicines including vaccination be administered at such intervals as may be decided by the authorised veterinary officers.

(29) The zoo operators shall arrange for medical check-ups of the staff responsible for upkeep of animals at least once in every six months to ensure that they do not have infections of such diseases that can infect the zoo animals.

(30) Each zoo shall maintain animal history sheets and treatment cards in respect of each animal of endangered species, identified by the Central Zoo Authority.

Verterinary Facilities:

(31) Every large and medium zoo shall have full-fledged veterinary facilities including a properly equipped veterinary hospital, basic diagnostic facilities and comprehensie range of drugs. Each veterinary hospital shall have isolation and quarantine wards for newly-ariving animals and sick animals. These wards should be so located as to minimise the chances of infections spreading to other animals of the zoo.

(32) Each veterinary hospital shall have facilities for restraining and handling sick animals including tranquilizing equipments and syringe projector. The hospital shall also have a reference library on animal health care and upkeep.

(33) The small and mini zoos, where full-fledged veterinary hospital is not available, shall have at least a treatment room in the premises of the zoo where routine examination of animals can be undertaken and immediate treatment can be provided.

(34) Every zoo shall have a post-mortem room. Any animal that dies in a zoo shall be subjected to a detailed post-mortem. and the findings recorded and maintained for a period of at least six years.

(35) Each zoo shall have a graveyard where the carcasses of dead animals can be buried without affecting the hygiene

and the cleanliness of the zoo. The large and medium zoos shall have an inscinerator for disposal of the carcasses and other refuse material.

Breeding of Animals:

(36) Every zoo shall formulate a programme for captive breeding of only such animals as are approved by the Central Zoo Authority for that zoo. They shall abide by the guidelines and directives of the Central Zoo Authority in this regad.

(37) Every zoo shall keep the animals in viable, social groups. No animal will be kept without a mate for a period exceeding one year unlss there is a legitimate reason for doing so or if the animal has already passed its prime and is of no use for breeding purposes. In the event of a zoo failing to find a mate for any single animal within this period, the animal shall be shifted to some other place according to the directions of the Central Zoo Authority.

(38) No zoo shall be allowed to acquire a single animal of any variety except when doing so is essential either for finding a mate for the single animal housed in the said zoo or for exchange of blood in a captive breeding group.

(39) Every zoo shall take up regular exchange programmes of animals so as to prevent the traits or ill effects of inbreeding. To achieve this objective each zoo shall maintain a stud book in respect of every endangered species.

(40) To safeguard against uncontrolled growth in the population of prolifically breeding animals, every zoo shall implement appropriate population control measures like separation of sexes, sterilizaiton, vasectomy, tubectomy and implanting of pallets etc.

(41) No zoo shall permit hybridization either between different species of animals or different races of the same species of animals. Maintenance of Records and Submission of Inventory to The Central Zoo Authority:

(42) Every zoo shall keep a record of the birth, acquisitions, sales, disposals and deaths of all animals. Th inventory of the

animals housed in each zoo as on 31st March of every year shall be submitted to the Central Zoo Authority by 30th April of the same year.

(43) Every zoo shall also submit a brief summary of the death of animals in the zoo for every financial year, alongwith the reasons of death identified on the basis of post-mortem reports and other diagnostic tests, by 30th April of the following year.

(44) Every zoo shall publish an annual report of the activities of the zoo in respect of each financial year. The copy of the said annual report shall be made available to the Central Zoo Authority, within two months, after the end of the financial year. The report shall also be made available to the general public at a reasonable cost.

Education and Research:

(45) Every enclosure in a zoo shall bear a sign board displaying scientific information regarding the animals exhibited in it.

(46) Every zoo shall publish leaflets, brochures and guidebooks and make the same available to the visitors, either free of cost or at a reasonable price.

(47) Every large and medium zoo shall make arrangements for recording, in writing, the detailed observations. about the biological behaviour, population dynamics and veterinary care of the animals exhibited as per directions of the Central Zoo Authority so that a detailed database could be developed. The database shall be exchanged with other zoos as well as the Central Zoo Authority.

Visitor Facilities:

(48) The zoo operators shall provide adequate civic facilities like toilets, visitor sheds, and drinking water points at convenient places in the zoo for visitors.

(49) First-aid equipments including anti-venom shall be readily available in the premises of the zoo.

(50) Arrangements shall be made to provide access to the zoo to disabled visitors including those in the wheel chair.

Development and Planning:

(51) Each zoo shall prepare a long-term master plan for its development. The zoo shall also prepare a management plan, giving details of the proposal and activities of development for next six years. The copies of the said plans shall be sent to the Central Zoo Authority.

APPLICATION FOR GETTING RECOGNITION FROM THE CENTRAL ZOO AUTHORITY

UNDER SECTION 38H (SUB-SECTION 2)

FORM - A

To

The Member-Secretary,

Central Zoo Authority of India,

NEW DELHI.

We want to get recognition under section 38H of the Wild Life (Protection) Act, 1972 in respect of ________________ Bank draft/Postal Order for Rs.500/- drawn in favour of Central Zoo Authority is also enclosed. The required information in respect of____________________is as under:

1. Name of the Zoo:

2. Location of the Zoo and Area:

3. Date of establishment:

4. Name of controling authority/operator:

5. Total number of visitors to the Zoo during the last three years (Year wise)

6. Total number of days on which zoo is open to visitors during a calender year.

7. Number of animals exhibited by the zoo:

Stock position during the current financial year

Number of Species exhibited	Stock Position Disposals on the close year	Births Stock as on of preceding	Acquitions	Deaths the date of application

MAMMALS
BIRDS
REPTILES
AMPHIBIANS
FISHES AND OTHERS
INVERTEBRATES

8. Total number of enclosures:

(i) Open air moated encloures:

(ii) Closed cages/aviaries:

9. List of endangered species bred during last 3 years

10. Veterinary facilities

(a) Whole time veterinarian available or not:

(b) Facilities available in the Veterinary Hospital:

(a) Operation theatre /Surgical room

(b) X-ray facility

(c) Squeeze cages

(d) In-door patient ward

(e) Quarantine ward

(f) Dispensary

(g) Nursery for hand-rearing animal babies

(h) Pathological laboratory

(i) Tranquilising equipments./drugs

11. Whether the following facilities exist in the zoo:

i) Kitchen

ii) Food store

iii) Deep Freeze

iv) Potable water facility

v) Food distribution van/rickshaw etc.

12. Sanitary care and disease control:

Whether –

i) Pollution free water to nimals for drinkng is available ?

ii) Proper drainage system exists in enclosures ?

iii) Regular disposal of refuse material is done ?

iv) Programme for control of pests and predators exists ?

v) Preventive measures like deworming and vaccination are being provided ?

13. Amenities to visitors:

Whether

(a) Public facilities like toilets /bathrooms exist ?

(b) Sufficient number of drinking water taps available ?

(c) Visitor information centre and nature interpretation centre exist ?

(d) Zoo eduation facilities have been provided ?

(e) Public telephone booths are available ?

(f) Kiosks and restaurants are available at the zoo ?

14. Safety measures for visitors:

Whether

(a) Effective stand-of barriers have been provided around enclosures ?

(c) Adequate number of warning sign boards exist ?

(d) First-Aid measures are availabe ?

15. Budget of the Zoo for last 3 years

Revenue Grants Total expenditure

16. Annual Report, Guide books, Brochure or any other publication (copies enclosed)

17. Master plan of the Zoo (copy enclosed)

Signature of the Applicant

MINIMUM PRESCRIBED SIZE FOR FEEDING/ RETIRING CUBICLE/ENCLOSURES FOR IMPORTANT MAMMALIAN SPECIES OF CAPTIVE ANIMALS.

Name of the Species.	Size of the cubicle/enclosures - in metres		
	length	breadth	height
FAMILY - Felidae:			
Tiger and lions	2.75	1.8	3
Panther	2	1.5	2
Clouded leopard & snow leopard	2	1.5	2
Small cats	1.8	1.5	1.5
FAMILY - Elephantidae			
Elephant.	8	6	5.5
FAMILY - Rhinocerotidae			
One-horned Indian Rhinoceros	5	3	2.5
FAMILY - Cervidae			
Brow antlered deer	3	2	2.5
Hangul	3	2	2.5
Swamp deer	3	2	2.5
Musk deer	2.5	1.5	2
Mouse deer	1.5	1	1.5
FAMILY - Bovidae			
Nilgiri tahr	2.5	1.5	2
Chinkara	2.5	1.5	2
Four horned antelope	2.5	1.5	2
Wild Buffalo	3	1.5	2
Indian Bison	3	2	2.5
Yak	4	2	2.5
Bharal, goral, wild sheep and markhor	2.5	1.5	2
FAMILY - Equidae			
Wild Ass	4	2	2.5
FAMILY - Ursidae			
All types of Indian bears.	2.5	1.8	2
FAMILY - Canidae			
Jackal, wolf & wild dog	2	1.5	1.5
FAMILY - Vivirridae			
Palm civet	2	1	1
Large Indian civet & binturong	2	1.5	1
FAMILY - Mustellidae			
Otters all types	2.5	1.5	1
Ratel/Hogbadger	2.5	1.5	1
Martens	2	1.5	1
FAMILY - Procyonidae			
Red Panda	3	1.5	1
FAMILY - Lorisidae			
Slow loris and slender loris	1	1	1.5
FAMILY-Cercopithecidae			
Monkeys and langurs.	2	1	1.5

NUMBER OF ZOOS AND CAPTIVE WILDLIFE FACILITIES IN STATES AND UNION TERRITORIES OF INDIA

STATE/	ZOOS NATURE/ PARKS	DEER AQUARIUMS PARKS	SAFARI EDUCATION PARKS	TOTAL BREEDING CENTRES		SNAKE	
Andaman and Nicobar Islands	1	0	0	0	0	0	1
Andhra Pradesh	3	14	3	0	1	1	22
Arunachal Pradesh	3	0	0	0	1	0	4
Assam	1	0	0	0	3	0	4
Bihar	5	1	0	0	0	0	6
Delhi	1	1	0	0	0	0	2
Goa	1	0	0	0	0	0	1
Dadra and Nagar Haveli (UT)	1	2	0	0	0	0	3
Gujarat	8	4	0	0	2	2	16
Haryana	5	2	0	0	1	0	8
Himachal Pradesh	4	1	2	0	3	0	10
Jammu & Kashmir	2	1	0	0	0	0	3
Karnataka	19	3	4	1	0	0	27
Kerala	3	1	1	1	2	0	8
Madhya Pradesh	5	0	0	0	1	1	7
Maharashtra	10	1	1	2	2	1	17
Manipur	1	0	0	0	0	0	1
Meghalaya	2	0	0	0	0	0	2
Mizoram	1	0	0	0	0	0	1
Nagaland	1	0	0	0	0	0	1
Orissa	2	7	3	0	1	0	13
Pondicherry	1	0	0	0	0	0	1
Punjab	5	3	1	0	0	0	9
Rajasthan	6	1	0	0	1	0	8
TamilNadu	8	1	0	1	1	2	13
Tripura	1	0	0	0	0	0	1
Sikkim	1	1	0	0	0	0	2
Uttar Pradesh	3	7	0	0	6	0	16
West Bengal	3	0	1	1	0	1	6
Total	107	*51	16	6	25	8	213

* The actual number could be much more but State-wise details are not available.

5

IMPORT-EXPORT POLICY

EXTRACTS FROM THE EXPORT & IMPORT POLICY U APRIL 1992-31 MARCH 1997) AS APPLICATION FROM 1.4.94 TO 31.3.96

Para 7: Definitions

(2) "Act" means the Foreign Trade (Development & Regulation) Act 1992 (No. 22 of 199.2).

(24) "Manufacture" means to make, produce, fabricate, assemble, process or bring into existenc, by hand or by machine, a new produce having a distinctive name, character or use and shall include process such as refrigeration, repacking, polishing labelling and segregation. Manufacture, for the purpose of this Policy, shall also include agriculture_ aquacultures, animal husbary floriculture, horticulture, pisciculture, poultry and sericullure.

(31) "Policy" means the Export & Import Policy 1992-97 as amended from time to time.

Para 8. Exports & Imports free unless regulated

Exports and imports may be done freely, except to the extent they are regulated by the provisions of this Policy or any other law for the time being in force.

Para 9. Form of Regulation

The Central Government may, in public interest, regulate the import or export of goods by means of a Negative list of Imports or a Negative List of Exports, as the case may be.

Para 10. Negative Lists

The Negative lists may consists of goods the import or export of which is prohibited, restricted through licensing or otherwise, or canalised. The Negative List of Imports and the Negative list of Exports shall be as contained in this Policy.

Para 11. Prohibited goods

Prohibited goods shall not be imported or exported.

Para 16. Procedure

The Director General of Foreign Trade may, in any case or class of cases, specify the procedure to be followed by an exporter or importer or by any licensing, competent or the authority for the purpose of implementing the provisions of the Act, the Rules and Orders made thereunder the this Policy. Such procedures shall be included in the Handbook of Procedures and published by means of a Public Notice. Such procedures may, in like manner, be amended from time to time.

NEGATIVE LIST OF IMPORTS PROHIBITED ITEMS

3. Wild animals including their parts and products a ivory.

RESTRICTED ITEMS SEEDS, PLANTS AND ANIMALS

1. Animals, Import permitted against a licence to zoos and zoological parks, recognised scientific/

Birds & research institutions, circus companies, private individuals, on the recommendation of Reptiles the Chief Wildlife Warden of a State Government subject to the provisions of the (including Convention on International Trade in Endangered Species of Wild Fauna & theirparts & Flora (CITES) products)

4. Plants,

(a) Import of seeds of wheat, paddy coarse cereals, pulses, oilseeds and fodder for Fruits & sowing is permitted without a licence subject to fulfilment of the provisions of the New Seeds Policy on Seed Development 1988 and in accordance with a permit for import granted under the Plants, Fruits and Seeds (Regulation of Import into India) Order 1989.

(b) Import of seeds of vegetables flowers, fruits and plants, tubers and bulbs of flowers, cutting, saplin& budwood etc.of flowers and fruits for sowing or planting is permitted without a licence in accordance with a permit for import granted under the plants, Fruits and Seeds (Regulation of Import into India) Order 1989.

(c) Import of Seeds, fruits and plants for consumption or other purpose is permitted against a licence or in accordance with Public Notice in this behalf.

(d) Import of plants, their products and derivatives shall also be subject to the provisions of the Convention on International Trade in Endangered Species of Wild Fanua and Flora (CITES).

NEGATIVE LIST OF EXPORTS PROHIBITED ITEMS

1. All forms of wild animals including their parts and products except Peacock tails including handicrafts made thereof and manufactured Articles and Shavings of Shed Antlers of Chital and Sambhar subject to condition as specified in Annexure to Public Notice No. 15-ETC (N)/92/97 dated 31st March, 1993 issued by the Director General of Foreign Trade and reproduced in the Handbook of Procedures. (Vol. 1)

2. Exotic Birds

3. All items of plants included in Appendix I of the Convention on International Trade in Endangered Species (CITES), wild orchids as well as plants as specified in Public Notice No. 47 (PN)/92-97 dated 30th March, 1994 issued by the Director General of Foreign Trade and reproduced in the Handbook of Procedures (Vol.1)

7. Wood and wood products in the form of logs, timber, stumps, roots, barks, chips, powder, flakes, dust, pulp and charcoal except sawn timber made exclusively out of imported teak logs/timber subject to conditions as specified in Annexures to Public Notice No. 15-ETC (PN)-92/97 dated 31st March, 1993 issued by the Director General of Foreign Trade and reproduced in the Handbook of Procedures (Vol.1)

9. Sandalwood.

10. Red Sanders wood in any form whether raw, processed or unprocessed as well as any product made thereof.

AESTRICTED ITEMS

(EXPORTS PERMITTED UNDER LICENCE)

8. Fur of domestic animals, excluding lamb fur skin.

10. Hides and skins, namely:

(i) Cuttings and fleshing of hides and skins used as raw materials for manufacture of animal glue gelatine.

(ii) Raw hides and skins, all types excluding lamb fur skin.

(iii) All categories of semi-processed hides and skins including E.I. tanned and wet blue hides and skins and crust leather.

(iv) clothing leather fur suede/hair, hair-on suede/shearing suede leathers.

(v) Fur leathers.

20. Seeds and planting materials namely:

Castor seeds, cotton seeds except such cotton seeds are of varieties hybrids of other countries, grown under custom production; cashew seeds and plants; Egyptial clover (Barseem) Trifloium alastum seeds, Fodder crop seeds; Green manure seeds other than Dhanincha; guar seeds (whole); jute seeds; linseeds; lucrene (alfalfa) medicago sative; mesta seeds; Nux vomica seeds /bark/leave/roots and powder thereof; onion seeds, seeds of ornamental plants (wild variety); paddy seeds (wild variety); pepper cuttings or rooted cuttings of pepper; persian clover (snaftel trifolium, - resupinatum) seeds; red sanders' seeds (Pterocarpus santalinus; rubber seeds; russa grass seeds and tufts; seeds of all forestry species; seeds of all oilseeds and pulses; soyabean seeds; sandalwod seeds (Santalum album); saffron seeds of corns (planting material for saffron); wheat seeds (wild variety).

21. Sea shells, excluding polished sea shells and handicrafts made out of sea shells of all species except those of the

undermentioned species the export of which shall not be allowed in any form.

i. Trochus niloticos

ii. Trubo species

iii Lambis species

iv. Tridacna gigas

v. Zancus pyrus

22. Sea weeds of all types including G. edulis but excluding brown sea weeds and agarophytes of Tamil Nadu coast origin processed form.

2. The Director General of foreign Tradc hereby porhibits the export of plants plant portions and their derviatives and extracts obtained from the wild as under:

1. Aconitum species
2. Attropa species
3. Aristolochia species
4. Angiopteris species
5. Arundinaria Jaunsarensia.
6. Balanophora species
7. Colchisum luteum (Hirantutya).
8. Commiphora whight ii.
9. Coptis species
10. Drosera species
11. Gentiana Kurroo (Kuru, Kutki).
12. Gloriasa superba.
13. Gnetum species
14. Iphignia indica.
15. Meconipsis betonicifolia.
16. Nardostachys species (Jatamansi)

17. Osmunda species
18. Rhododendron species
19. Physochlaina praealta (Bajarbang)
20. Praltia serpumlia
21. Rheum emodi (Dolu)
22. Berberis aristata (Indian barberry; Rasvat)
23. Acorus species
24. Artemisia species
25. Coscinium fenestratum (Calumba wood).
26. Costus speciousas (Kew, Kust).
27. Didymocarpus pedicellata.
28. Dolomiaea pedicellata
29. Ephedra species.
30. Gynocardia odorata (Chaulmogri).
31. Hydnocarpus species.
32. Hyoscymus niger (Broseword).
33. Strychnos potatorum (Nirmali).
34. Swertia chirata (Charayatah).
35. Urgines species.
36. Beddomes cycad (Cycas beddomei).
37. Blue vanda (Vandaeoerules).
38. Kuth (Saussurea lappa).
39. Ladies slipper orchid (Paphiopedilium species).
40. Pitcher plant (Nepenthes Khasiana).
41. Red vanda (Renenthera imschooiana).
42. Rauvolifia serpentina (Sarpagandha).
43. Ceropegia species.
44. Frerea indica (Shindal Mankundi).

45. Araucaria araucana (Monkey-puzzle tree).

46. Podophyllum hexandurm (emodi) Indian Podophyllum).

47. Cactacea species (Cactus).

48. Cyatheaceae species (Tree Ferns).

49. Cycadacea species (Cycads).

50. Dioscorea deltoidea (Elephant's Foot).

51. Euphorbia species (Euphorbias).

52. Aloe species (Aloes)

53. Orchidaceae species (Orchids)

54. Pterocarpus Santalinus (Red Sanders).

55. Taxus wallichiana (Common Yew of Birmi leaves).

56. Aquilaria malaccensis (Agarwood).

3. Plants and plant portions, derivatives and extracts (including value added herba formulations) of the cultivated varieties of the species above (excluding serial no.54) will be allowed for export subject to production of a certificate of cultivation from Regional Deputy Director (Wildlife), or Chief conservator of Forests or Divisional Forest Officers of the state concerned from where these plants and plant portions have been procured. However in respect of cultivated varieties of the species covered by Appendix I (S.No. 36 to 41 of Paragraph 2 above) and Appendix II (s. No. 42 to 56 of Paragraph 2 above) of CITES, a CITES Permit for export will also be required.

4. Exports allowed only through six major ports viz. Bombay, Calcutta, Cochin, Delhi, Madras, and Tuticorin.

5. This issue is public interest.

THE WILDLIFE (PROTECTION) LICENSING (ADDITIONAL MATTERS FOR CONSIDERATION) RULES, 1983

G.S.R. 328 (E), DATED 13TH APRIL, 19831

In exercise of the powers conferred by Cf. (a) of sub-section (1) of sec. 63, read with Cf. (b) of sub-section (4) of sec. 44 of

Wildlife (Protection) Act, 1972 (53 of 1972) the Central Government hereby makes the following rules, namely

1. Short title, extent, and commencement

(1) These rules may be called the Wildlife (Protection) Licensing (Additional Matters for Consideration) Rules, 1983

(2) They shall extend to the whole of India except the State of Jammu and Kashmir.

(3) They shall come into force on the date of their publication in the official Gazette.

2. Definition

In these Rues, unless the context otherwise, requires, "Act" means the Wildlife (Protection) Act, 1972 (53 of 1972).

3. Additional matters for consideration for grant of licence under sec. 44 of the Act.

For the purposes of granting a licence referred to in sub-section (1) of Sec 44 of the Act, the Chief Wildlife Warden or the authorised officer, as the case may be, shall in addition to the matters specified in Cf. (b) of sub-section (4) of that section, have regard to the following other matters, namely

(i) capacity of the applicant to handle the business concerned with reference to facilities, equipment, and suitability of the premises for such business;

(ii) the source and the manner in which the supplies for the business concerned would be obtained;

(iii) number of licences for the relevant business already in existence in the area concerned;

(iv) implications which the grant of such licence would have on the hunting or trade of the wild animals concerned.

Provided that no such shall be granted if the said implications relate to any wild animal specified in Sch.I or part 11 of Sch.II to the Act, except with the previous consultation of the Central Government.

THE WILDLIFE (TRANSACTION AND TAXIDERMY) RULES, 1973

G.S.R. 198(E), DATED THE 9TH APRIL, 1973

In exercise of the powers conferred by Cl.(b) of sub-section (1) of Sec.63 of the Wildlife (Protection) Act, 1972 (53 of 1972), the Central Government hereby makes the following rules.

(1) Short title, extent, commencement

(1) These rules may be called the Wildlife (Transactions and Taxidermy) Rules, 1973.

(2) They extend to the whole of the State of Bihar, Gujarat, Haryana, Himachal Pradesh, Madhya Pradesh and Uttar Pradesh.

(3) They shall come into force on the 9th April 1973.

2. Definition

In these rules, unless the context other-wise requires

(a) "Act" means the Wildlife (Protection) Act, 1972 (53 of 1972);

(b) "form" means a Form appended to these rules;

(c) "Licensee" means a licensee under Chapter V of the Act;

(d) "Officer" means the Chief Wildlife Warden or any other officer whom the State Government may, for the purposes of these rules, by notification in the official Gazette, appoint;

(e) "specified animal" means any animal which is specified in Sch.I or Part II of Sch.II to the Act and which is

(i) captured or kept or bred in captivity, or

(ii) found wild in nature.

3. Acquiring, receiving, or keeping specified animal, etc. in control, custody, or possession or put under process of taxidermy or make articles, etc.,

(1) No person shall

(i) acquire, receive, keep in his control, custody or possession, any specified animal or any animal articles, trophy, uncured trophy, or meat derived therefrom, or

(ii) put under a process of taxidermy or make animal articles containing part of whole of such animal, except with the previous permission of the officer.

(2) Every application for such permission shall be made in Form 1.

(3) On receipt of an application made under sub-rule (2), the officer may, after making such inquiry as he may think fit and within a period of fifteen days from the date of receipt of the application, either grant or refuse to grant the permission;

Provided that no such permission shall be granted unless the officer is satisfied that the specified animal or animal article, trophy, uncured trophy, or meat, referred to in sub-rule (1) has been lawfully acquired.

(4) Where the officer refuses to grant the permission, he shall record the reason for so doing and a copy of the reason so recorded shall be communicated to the licensee applying for the permission.

(5) Every permission granted under sub-rule (3) shall be in Form II.

4. Submission of report of stocks

(1) Every licensee to whom permission has been granted under sub-rule (3), of Rule 3 shall submit, to the officer who has granted the said permission, report regarding the stocks of specified animal or animal article, trophy, uncured trophy, or meat, referred to in sub-rule (1) of Rule 3, in Form III within a period of [thirty days] of the acquisition, receipt or keeping of the same in his control, custody or possession.

(2) The officer, after receiving such report, may arrange to affix identification marks on such stocks.

5. Sale of specified animal. etc.

(1) No licensed dealer shall sell or offer for sale any specified animal or any animal article, trophy, or uncured trophy derived therefrom, except to a person authorised to purchase by a permission granted by the officer and where the sale is effected

the purchaser shall surrender the permission to the licensed dealer.

(2) Every application for permission to purchase shall be made in Form IV.

(3) On receipt of an application made under sub-rule (2), the officer may, after making such inquiry as he may think fir, and within a period of ten days from the date of receipt of the application, either grant or refuse to grant the permission.

(4) Where the officer refuses to grant the permission, he shall record the reasons for so doing and a copy of the reasons so recorded shall be communicated to the person applying for the permission.

(5) Every permission granted under sub-rule (3) shall be valid up to a period of one month from the date of issue of the same.

(7) Every licensed dealer shall, at the time of each sale, issue a voucher in relation to the specified animal or animal article, trophy, or uncured trophy referred to in sub-rule (1), to the person authorised to purchase.

(8) Each voucher shall contain the following particulars, namely

(a) date of issue of the voucher;

(b) the amount of price realised or to be realised;

(c) name and address of the licensed dealer issuing the voucher;

(d) name and address of person to whom the voucher is issued;

(e) permission number of the person authorised to purchase;

(f) description of the specified animal/animal article /trophy /uncured trophy derived therefrom and number;

(g) whether such specified animal/animal article/trophy / uncured trophy was/were required to be declared under Sec.44 of the Wildlife (Protection) Act, 1972 (53 of 1972), and if so, whether it/they has /have been declared;

(h) signature of the licensed dealer issuing the voucher;

(i) signature of the person to whom the voucher is issued

6. Taxidermy or making animal article

(1) Every licensed taxidermist or licensed manufacturer shall, at the time of returning the trophy or animal article, issue a voucher to the owner of the said trophy or animal article.

(2) Each voucher shall contain the following particulars, namely

(a) date of issue of voucher;

(b) charges realised or to be realised;

(c) name and address or the licensed taxidermist/ manufacturer issuing the voucher;

(d) name and address of the person to whom the voucher is issued;

(f) whether uncured trophy/ trophy/animal article was required to be declared under Sec.40. or Sec. 44 of the Wildlife (Protection) Act, 1972 (53 of 1972), and if so whether it/they has/have been declared;

(g) signature of the licensed taxidermist / manufacturer issuing the voucher.

7. Maintenance of vouchers

(1) The voucher referred to in Rules 5 or 6 shall be in triplicate and serially numbered.

(2) The duplicate and the triplicate copy of the voucher shall be retained by the licensed dealer" licensed taxidermist, or licensed manufacturer, and the original copy of the voucher shall be given to the person referred to in sub-rule (7) of Rule 5 or sub-rule (1) of Rule 6.

(3) Every book containing blank vouchers shall be presented to the officer for affixing his initials to stamps on such book before it is brought into use.

(4) (a) Every licensed dealer, licensed taxidermist, or licensed manufacturer shall send in monthly batches, not later than the seventh day of every month, the duplicate copies of vouchers retained by him, to the officer.

(b) Every permission surrendered to a licensed dealer at the time of sale shall also be enclosed along with the duplicate copies aforesaid.

8. Transport of specified animal. etc.

(1) No licensee shall transport from one place to another within the State any specified animal, animal articles, trophy, or uncured trophy derived therefrom, except with the previous permission of the officer.

(2) Every application for such permission shall be made in Form VI.

(3) On receipt of an application made under sub-rule (2) the officer may, after making such inquiry as he may think fit, and within a period of seven days from the date of receipt of the application, either grant or refuse to grant the permission.

Provided that no such permission shall be granted unless the officer is satisfied that the specified animal, or animal article, trophy or uncured trophy, referred to in sub-rule (1) has been lawfully acquired.

(4) Where the officer refuses to grant the permission, he shall record the reasons for so doing and a copy of the reasons so recorded shall be communicated to the licensee applying for the permission.

(5) Every permission granted under sub-rule(3) shall be in Form VII.

9. Appeal

(1) Any licensee or a person aggrieved by an order made by the Chief Wildlife Warden or any officer granting the permission under sub-rule (3) of Rule 3, sub-rule (3) of Rule 5, of sub-rule (3) of Rule 8, may prefer an appeal

(i) if the order is made by an officer other than the Chief Wildlife Warden, to the Chief Wildlife Warden, or

(ii) if the order is made by the Chief Wildlife Warden , to the State Government.

(2) In the case of an order passed in appeal by the Chief Wildlife Warden under Cl (i) of sub-rule (1), a second appeal shall lie to the State Government.

(3) No appeal shall be entertained unless it is preferred within fifteen days from the date of the communication to the applicant of the order appealed against.

Provided that the appellate authority may admit any appeal after the expiry of the period aforesaid, if it is satisfied that the appellant had sufficient cause for not preferring the appeal in time.

GUIDELINES FOR APPOINTMENT OF HONORARY WILDLIFE WARDENS

The Need

1. People's participation and support is crucial for nature and Wildlife conservation, One of the important ways of enlisting such support is by involving the community leaders and other persons of standing, who have the interest as well as the capacity to render assistance for this cause. Such assistance can be very useful in control over poaching for this clandestine trade in wild animals or their articles, identification of relatively less known wildlife refuges needing protection, carrying the message of conservation to the people living in and around the sanctuaries and national parks, and related matters. This objective can be accomplished if really suitable public men are identified, duties and appointed Honorary Wildlife Wardens, with their responsibilities, and powers clearly defined.

Legal Status

2. Section 4 of the Wildlife (Protection) Act. 1972 empowers the State Govermnent to appoint

(a) a Chief Wildlife Warden

(b) Wildlife Wardens; and

(c) such other officers and employees as may be necessary for the purposes of the Act.

Honorary Wildlife Wardens can be appointed under sub-section (c) of section 4 of the Act. Under section 59 of the aforesaid Act, such Honorary Wildlife Wardens shall be deemed to be public servants within the meaning of section 21 of the Indian Penal Code.

Criteria for Selection

3. It is very important that the right persons are selected for appointment as Honorary Wildlife Wardens. Every State has a Chief Wildlife Warden and it is mainly his duty to recommend the names of suitable person for this purpose. However, in order to assist him in this regard as well as to introduce a measure of wider participation, the members of the State Wildlife Advisory Board should be requested to suggest suitable names, especially from their own areas.

4. The following criteria should be kept in mind while assessing the suitability of a person as an Honorary Wildlife Warden:

(a) Genuine concern for Wildlife conservation.

(b) Personal record free of involvement in any activity detrimental to the interests of nature and wildlife conservation.

Any person involved in commercial exploitation of Wildlife should not be considered.

(c) Capacity to render help to the official machinery.

(d) Local standing which make him/her effective, especially in conveying the conservation message.

5. An important point to bear in mind is the identification of areas particularly prone to poaching, e.g. forests in the vicinity of urban centres and cantonments or close to sanctuaries and national park. Likewise, centres of clandestine trade in wildlife and products thereof should be identified; so also areas where damage to people or their property from wild animals is heavy. Selection of persons as Honorary Wildlife Wardens

must be related to such problem areas because it is these areas which need priority attention and where public participation is needed most.

Procedure for Appointment

Under Section 6 of the Wildlife (Protection) Act, 1972, every State and Union Territory has a Wildlife Advisory Board to aid and advise the Government in matters connected with the protection of wildlife. The appointment of an Honorary Wildlife Warden should be generally with the recommendation of this Advisory Board. The Chief Wildlife Warden should submit the proposals for this purpose at the meetings (s) of the Board and then seek the orders of the Government.

While recommending any person for such appointment, the criteria laid down in paras 4 and 5 above must be kept in mind by the Board. The appointment of an Honorary Wildlife Warden should, in the first instance, be generally for a period of one year. Thereafter on the recommendation of the Wildlife Advisory Board, it may be renewed for a period not exceeding 2-3 years at a time.

The Wildlife Advisory Board of each State/Union Territory should review the functioning of the scheme of Honorary Wildlife Wardens at least once every year. The appointment order of an Honorary Wildlife Warden should clearly specify the jurisdiction, which should normally be a district or a few districts, in the area where the person resides. However, there is no objection to making members of the State Wildlife Advisory Board Honorary Wardens for larger areas.

Each Honorary Warden should be issued an identity Card having his signature and photograph duly attested by the Chief Wildlife Warden. The Chief Wildlife Warden should also give each Honorary Warden a small booklet containing the Wildlife (Protection) Act and the Rules made thereunder as well as the duties, responsibilities, and power of an Honorary Wildlife Warden.

The State Government may, at its discretion, terminate the appointment of an Honorary Wildlife Warden at any time, without assigning reasons.

Duties and Responsibilities

13. The main duty and responsibility of an Honorary Wildlife Warden is to assist whole heatedly the State organization responsible for wildlife conservation work, especially with regard to the following matters :

(a) Control of poaching and clandestine trade in wild animals and products / articles thereof

(b) Detection and prosecution of offences under the Wildlife (Protection) Act and the Rules made thereunder.

(c) Preventing damage to the habitat of wildlife.

(d) Identification and selection of areas suitable to be declared as sanctuaries, national parks, closed areas, etc; as well as measures for their proper protection.

(e) Measures for dealing with the problem of damage by wild animals to life and property, including the assessment and payment of compensation, etc.

(f) Carrying the message of conservation to the people and enlisting public support for nature and wildlife conservation. The effort should be specially directed towards the communities living in or near the declared wildlife reserves.

(g) Any other matter connected with the protection of wildlife, which may be entrusted by the Wildlife Advisory Board or the Chief Wildlife Warden of the State, from time to time.

Powers

14. In accordance with sub-section (3) of section 4 of the Wildlife (Protection) Act, 1M, an Honorary Wildlife Warden appointed under sub-section (2) (c) of section 4 shall be subordinate to the Chief Wildlife Warden of the State and under section 59 of aforesaid Act, he shall be deemed to be a public servant within the meaning of section 21 of the Indian Penal Code. Protection for action taken in good faith is provided under section 60 of the Act.

15. With a view of making the Honorary Wildlife Wardens useful and effective it is necessary that the following specific

powers under the Wildlife (Protection) Act, 1972 should be delegated to them:

(a) Power to inspect records of licences under section 47 (b) of the Act;

(b) Powers of entry, search, seizure, and detention under section 50 for prevention and detection of offences under the Act.

16. Suitable Honorary Wildlife Wardens could be authorised also to file complaints in courts in accordance with section 55 of the Wildlife (Protection) Act, 1972. Normally, however, an Honorary Wildlife Warden should bring the offence detected by him to the notice of the Wildlife Warden having jurisdiction for making proper investigation and lodging a complaint in the Court as laid down in section 55 of the Act.

17. Apart from the above, the State Government may delegate any other power under the aforesaid Act, as it may consider necessary.

General

18. Just as it is expected that the Honorary Wildlife Wardens should assist the State Wildlife organization, it is equally essential that the Chief Wildlife Warden and the whole State Machinery responsible for the protection of wildlife should take all possible steps to associate the Honorary Wildlife Wardens in their work. This can be achieved best by fostering a spirit of mutual trust and confidence.

19. No staff or vehicle support can be provided to Honorary Wardens as a matter of course. However, if the circumstances warrant, the departmental staff should provide all possible help and assistance. Instructions to this effect should be issued by the State Government to all concerned officers in the field.

20. It is also appropriate that the actual expenses incurred by an Honorary Warden on travel by public transport for carrying out the duties assigned to him should be reimbursed by the State Government. In addition, all actual expenses incurred in the detection of an offence under the Wildlife

(Protection) Act, which leads to successful prosecution may be reimbursed after due verification.

21. The State Government should recognise outstanding work or service rendered by any Honorary Warden. Such recognition can be by way of a letter of commendation, or a certificate signed by the Minister in charge of the Department, or the membership of the State Wildlife Advisory Board. Cash grants could also be considered in suitable cases.

FOREST (CONSERVATION) ACT, 1980 WITH AMENDMENTS MADE IN 1988

An Act to provide for the conservation of forests and for matters connected therewith or ancillary or incidental thereto.

Be it enacted by Parliament in the Thirty-first Year of the Republic of India as follows:-

1. Short title, extent and commencement.

(1) This Act may be called the Forest (Conservation) Act, 1980.

(2) It extends to the whole of India except the State of Jammu and Kashmir.

(3) It shall be deemed to have come into force on the 25th day of October, 1980.

2. Restriction on the dereservation of forests or use of forest land for non-forest purpose.

Notwithstanding anything contained in any other law for the time being in force in a State, no State Government or other authority shall make, except with the prior approval of the Central Government, any order directing-

(i) that any reserved forest (within the meaning of the expression "reserved forest" in any law for the time being in force in that State) or any portion thereof, shall cease to be reserved;

(ii) that any forest land or any portion thereof may be used for any non-forest purpose;

(iii) that any forest land or any portion thereof may be assigned by way of lease or otherwise to any private person or to any authority, corporation, agency or any other organisation not owned, managed or controlled by Government;

(iv) that any forest land or any portion thereof may be cleared of trees which have grown naturally in that land or portion, for the purpose of using it for reafforestation.

Explanation - For the porpose of this section, "non-forest purpose" means the breaking up or clearing of any forest land or portion thereof for-

(a) the cultivation of tea, coffee, spices, rubber, palms, oil-bearing plants, horticultural crops or medicinal plants;

(b) any purpose other than reafforestation; but does not include any work relating or ancillary to conservation, development and management of forests and wildlife, namely, the establishment of check-posts, fire lines, wireless communications and construction of fencing, bridges and culverts, dams, waterholes, trench marks, boundary marks, pipelines or other like purposes.

3. Constitution of Advisory Committee.

The Central Government may constitute a Committee consisting of such number of persons as h may deem fit to advise that Government with regard to-

(i) the grant of approval. under Section 2; and

(ii) any other matter connected with the conservation of forests which may be referred to h by the Central Government.

3A. Penalty for contravention of the provisions of the Act.

Whoever contravenes or abets the contravention of any of the provisions of Section 2, shall be punishable with simple imprisonment for a period which may extend to fifteen days.

3B. Offences by the Authorities and Government Departments.

(1) Where any offence under this Act has been committed-

(a) by any department of Government, the head of the department; or

(b) by any authority, every person who, at the time the offence was committed, was directly in charge of, and was responsible to, the authority for the conduct of the business of the authority as well as the authority; shall be deemed to be guilty of the offence and shall be liable to be proceeded against and punished accordingly:

Provided that nothing contained in this sub-section shall render the head of the department or any person referred to in clause (b), liable to any punishment if he proves that the offence was committed without his knowledge or that he exercised all due diligence to prevent the commission of such offence.

(2) Notwithstanding anything contained in sub-section (1), where an offence punishable under the Act has been committed by a department of Government or any authority referred to in clause (b) of sub-section (1) and it is proved that the offence has been committed with the consent or connivance of; or is attributable to any neglect on the part of any officer, other than the head of the department, or in the case of an authority, any person other than the persons referred to in clause (b) of sub-section (1), such officer or persons shall also be deemed to be guilty of that offence and shall be liable to be proceeded against and punished accordingly.

4. Power to make rules.

(1) The Central Government may, by notification in the Official Gazette, makes rules for carrying out the provisions of this Act.

(2) Every rule made under this Act shall be laid, as soon as may be after it is made, before each House of Parliament, while it is in session, for a total period of thirty days which may be comprised in one session or in two or more successive sessions, and if, before the expiry of the session immediately following the session or the successive sessions aforesaid, both Houses agree in making any modification in the rule or both Houses

agree that the rule should not be made, the rule shall thereafter have effect only in such modified form or be of no effect, as the case may be; so, however, that any such modification or annulment shall be without prejudice to the validity of anything previously done under that rule.

5. Repeal and saving.

(1) The Forest (Conservation) Ordinance, 1980 is hereby replaced.

(2) Notwithstanding such repeal, anything done or any action taken under the provisions of the said Ordinance shall be deemed to have been done or taken under the corresponding provisions of this Act.

6

THE INDIAN FOREST ACT

An Act to consolidate the law relating to forests, the transit of forest-produce and the duty leviable on timber and other forest-produce.

Whereas it is expedient to consolidate the law relating to forests, the transit of forest produce and the duty leviable on timber and other forest-produce; It is hereby enacted a follows:

PRELIMINARY

1. Short title and extent.–(1) This Act may be called the Indian Forest Act, 1921

1[(2) It extends to the whole of India except the territories which, immediately before the 1st November, 1956, were comprised in Part B States.

(3) It applies to the territories which, immediately before the 1st November, 1956, were comprised in the States of Bihar, Bombay, Coorg, Delhi, Madhya Pradesh, Orissa, Punj Uttar Pradesh and West Bengal; but the Government of any State may by notification in Official Gazette bring this Act into force2 in the whole or any specified part of that State which this Act extends and where it is not in force.]

2. Interpretation clause.–In this Act, unless there is anything repugnant in the subject or context–

(1) "cattle" includes elephants, camels, buffaloes, horses, mares, geldings, ponies colts, fillies, mules, asses, pigs, rams, ewes, sheep, lambs, goats and kids;

(2) "Forest-officer" means, any person whom 3 [* * *] the State Government or any office empowered by 3 [* * *] the State Government in this behalf, may appoint to carry out all any of the purposes of this Act or to do anything required by this Act or any rule m thereunder to be done by a Forest-officer;

(3) "forest-offence" means an offence punishable under this Actor under any rule made thereunder;

(4) "forest-produce" includes–

(a) the following whether found in, or brought from, a forest or not, that is to say timber, charcoal, caoutchouc, catechu, wood-oil, resin, natural varnish, bark, lac, mahua flowers, mahua seeds,4[kuth] and myrabolams, and

(b) the following when found in, or brought from a forest, that is to say

(i) trees and leaves, flowers and fruits, and all other parts or produce not herein before mentioned, of trees,

(ii) plants not being trees (including grass, creepers, reeds and moss), and all parts or produce of such plants,

(iii) wild animals and skins, tusks, horns, bones, silk, cocoons, honey and wax, and all other parts or produce of animals, and

(iv) peat, surface soil, rock and minerals (including limestone, laterite, mineral oils, and all products of mines or quaries);

(4A) "ownee" includes a Court of Wards in respect of property under the superintendence or charge of such Court;

(5) "river" includes any stream, canal, creek or other channels, natural or artificial;

(6) "timber" includes trees, when they have fallen or have been felled, and all wood whether cut up or fashioned or hollowed out for any purpose or not; and

(7) "tree" includes palms, bamboos, skumps, brush-wood and canes.

OF RESERVED FORESTS

3. Power to reserve forests.–The State Government may constitute any forest-land or waste-land which is the property

of Government, or over which the Government has proprietary rights, or to the whole or any part of the forest-produce of which the Government is entitled, a reserved forest in the manner hereinafter provided.

4. Notification by State Government.-

(1) Whenever it has been decided to constitute any land a reserved forest, the State Government shall issue a notification in the Official Gazette–

(a) declaring that it has been decided to constitute such land a reserved forest;

(b) specifying, as nearly as possible, the situation and limits of such land; and

(c) appointing an officer (hereinafter called "the Forest Settlement-officer") to inquire into and determine the existence, nature and extent of any rights alleged to exist in favour of any person in or over any land comprised within such limits or in or over any forest-produce, and to deal with the same as provided in this Chapter.

Explanation.–For the purpose of clause (b), it shall be sufficient to describe the limits of the forest by roads, rivers, ridges or other well-known or readily intelligible boundaries.

(2) The officer appointed under clause (c) of sub-section (1) shall ordinarily be a person not holding any forest-office except that of Forest Settlement-officer.

(3) Nothing in this section shall prevent the State Government from appointing any number of officers not exceeding three, not more than one of whom shall be a person holding any forest-office except as aforesaid, to perform the duties of a Forest Settlement-officer under this Act.

5. Bar of accrual of forest-rights.–After the issue of a notification under section 4, no right shall be acquired in or over the land comprised in such notification, except by succession or under a grant or contract in writing made or entered into by or on behalf of the Government or some person in whom such right was vested when the notification was issued; and

no fresh clearings for cultivation or for any other purpose shall be made in such land except in accordance with such rules as may be made by the State Government in this behalf.

6. Proclamation by Forest Settlement-officer.–When a notification has been issued under section 4, the Forest Settlement-officer shall publish in the local vernacular in every town and village in the neighbourhood of the land comprised therein, a proclamation

(a) specifying, as nearly as possible, the situation and limits of the proposed forest;

(b) explaining the consequences which, as hereinafter provided, will ensue on the reservation of such forest; and

(c) fixing a period of not less than three months from the date of such proclamation, and requiring every person claiming any right mentioned in section 4 or section, 5 within such period either to present to the Forest Settlement-officer a written notice specifying or to appear before him and state, the nature of such right and the amount and particulars of the compensation (if any) claimed in respect thereof.

7. Inquiry by Forest Settlement-officer.–The Forest Settlement-officer shall take down in writing all statements made under section 6, and shall at some convenient place inquire into all claims duly preferred under that section, and the existence of any rights mentioned in section 4 or section 5 and not claimed under section 6 so far as the same may be ascertainable from the records of Government and the evidence of any persons likely to be acquainted with the same.

8. Powers of Forest Settlement-officers.-For the purpose of such inquiry, the Forest Settlement-officer may exercise the following powers, that is to say:

(a) power to enter, by himself or any officer authorised by him for the purpose, upon any land, and to survey, demarcate and make a map of the same; and

(b) the powers of a Civil Court in the trial of suits.

9. Extinction of rights.-Rights in respect of which no claim has been preferred under section 6, and of the existence of

which no knowledge has been acquired by inquiry under section 7, shall be extinguished, unless before the notification under section 20 is published, the person claiming them satisfies the Forest Settlement-officer that he had sufficient cause for not prefer-ring such claim within the period fixed under section 6.

10. Treatment of claims relating to practice of shifting cultivation.—

(1) In the case of a claim relating to the practice of shifting cultivation, the Forest Settlement-officer shall record a statement setting forth the particulars of the claim and of any local rule or order under which the practice is allowed or regulated, and submit the statement to the State Government, together with his opinion as to whether the practice should be permitted or prohibited wholly or in part.

(2) On receipt of the statement and opinion, the State Government may make an order permitting or prohibiting the practice wholly or in part.

(3) If such practice is permitted wholly or in part, the Forest Settlement-officer may arrange for its exercise

(a) by altering the limits of the land under settlement so as to exclude land of sufficient extent, of a suitable kind, and in a locality reasonably convenient for the purposes of the claimants, or

(b) by causing certain portions of the land under settlement to be separately demarcated, and giving permission to the claimants to practise shifting cultivation therein under such conditions as he may prescribe.

(4) All arrangements made under sub-section (3) shall be subject to the previous sanction of the State Government.

(5) The practice of shifting cultivation shall in all cases be deemed a privilege subject to control, restriction and abolition by the State Government.

11. Power to acquire land over which right is claimed.–(1) In the case of a claim to a right in or over any land, other than a right of way or right of pasture, or a right to forest produce

or a water-course, the Forest Settlement-officer shall pass an order admitting or rejecting the same in whole or in part.

(2) If such claim is admitted in whole or in part, the Forest Settlement-officer shall either

(i) exclude such land- from the limits of the proposed forest; or

(ii) come to an agreement with the owner thereof for the surrender of his rights; or

(iii) proceed to acquire such land in the manner provided by the Land Acquisition Act, 1894 (1 of 1894).

(3) For the purpose of so acquiring such land

(a) the Forest Settlement-officer shall be deemed to be a Collector proceeding under the Land Acquisition Act, 1894 (1 of 1894);

(b) the claimant shall be deemed to be a person interested and appearing before him in pursuance of a notice given under section *9* of that Act;

(c) the provisions of the preceding sections of that Act shall be deemed to have been complied with; and

(d) the Collector, with the consent of the claimant, or the Court, with the consent of both parties, may award compensation in land, or partly in land and partly in money.

12. Order on claims to rights of pasture or to forest-produce.– In the case of a claim to rights of pasture or to forest-produce, the Forest Settlement-officer shall pass an order admitting or rejecting the same in whole or in part.

13. Record to be made by Forest Settlement-officer.–The Forest Settlement officer, when passing any order under section 12, shall record, so far as may be practicable,–

(a) the name, father's name, caste, residence and occupation of the person claiming the right; and

(b) the designation, position and area of all fields or groups fields (if any), and the designation and position of all buildings (if any) in respect of which the exercise of such rights is claimed.

14. Record where he admits claim.–

If the Forest Settlement-officer admits in whole or in part any claim under section 12, he shall also record the extent to which the claim is so admitted, specifying the number and description of the cattle which the claimant is from time to time entitled to graze in the forest, the season during which such pasture is permitted, the quantity of timber and other forest produce which he is from time to time authorised to take or receive, and such other particulars as the case may require. He shall also record whether the timber or other forest-produce obtained by the exercise of the rights claimed may be sold or bartered.

15. Exercise of rights admitted.-

(1) After making such record the Forest Settlement officer shall, to the best of his ability, having due regard to the maintenance of the reserved forest in respect of which the claim is made, pass such orders as will ensure the continued exercise of the rights so admitted.

(2) For this purpose the Forest Settlement-officer may

(a) set out some other forest-tract of sufficient extent, and in a locality reasonably convenient, for the purposes of such claimants, and record an order conferring upon them a right of pasture or to forest-produce (as the case may be) to the extent so admitted; or

(b) so alter the limits of the proposed forest as to exclude forest-land of sufficient extent, and in a locality reasonably convenient, for the purposes of the claimants; or

(c) record an order, continuing to such claimants a right of pasture or to forest-overpage produce, as the case may be, to the e tent so admitted, at such seasons, within such portions of the proposed forest, and under such rules, as may be made in this behalf by the State Government.

16. Commutation of rights.–In case the Forest Settlement-officer finds it impossible having due regard to the maintenance of the reserved forest, to make such settlement under section 15 as shall ensure the continued exercise of the said rights to

the extent so admitted, he shall, subject to such rules as the State Government may make in this behalf, commute such rights, by the payment to such persons of a sum of money in lieu thereof, or by the grant of land, or in such other manner as he thinks fit.

17. Appeal from order passed under section 11, section 12, section 15 or section 16.–Any person who has made a claim under this Act, or any Forest-officer or other person generally or specially empowered by the State Government in this behalf, may, within three months from the date of the order passed on such claim by the Forest Settlement-officer under section 11, section 12, section 15 or section 16, present an appeal from such order to such officer of the Revenue Department of rank not lower than that of a Collector, as the State Government may, by notification in the Official Gazette, appoint to hear appeals from such orders:

Provided that the State Government may establish a Court (hereinafter called the Forest Court) composed of three persons to be appointed by the State Government, and when the Forest Court has been so established, all such appeals shall be presented to it.

18. Appeal under section 17.–

(1) Every appeal under section 17 shall be made by petition in writing, and may be delivered to the Forest Settlement-officer, who shall forward it without delay to the authority competent to hear the same.

(2) If the appeal be to an officer appointed under section 17, it shall be heard in the manner prescribed for the time being for the hearing of appeals in matters relating to land-revenue.

(3) If the appeal be to the Forest Court, the Court shall fix a day and a convenient place in the neighbourhood of the proposed forest for hearing the appeal, and shall give notice thereof to the parties, and shall hear such appeal accordingly.

(4) The order passed on the appeal by such officer or Court, or by the majority of the members of such Court, as the case may be, shall, subject only to revision by the State Government, be final.

19. Pleaders.–

The State Government, or any person who has made a claim under this Act, may appoint any person to appear, plead and act on its or his behalf before the Forest Settlement-officer, or the appellate officer or Court, in the course of any inquiry or appeal under this Act.

20. Notification declaring forest reserved.–(1) When the following events have occurred, namely:–

(a) the period fixed under section 6 for preferring claims have elapsed and all claims (if any) made under that section or section 9 have been disposed of by the Forest Settlement-officer;

(b) if any such claims have been made, the period limited by section 17 for appealing from the orders passed on such claims has elapsed, and all appeals (if any) presented within such period have been disposed of by the appellate officer or Court; and

(c) all lands (if any) to be included in the proposed forest, which the Forest Settlement-officer has, under section 11, elected to acquire under the Land Acquisition Act, 1894 (1 of 1894), have become vested in the Government under section 16 of that Act, the State Government shall publish a notification in the Official Gazette, specifying definitely, according to boundary-marks erected or otherwise, the limits of the forest which is to be reserved, and declaring the same to be reserved from a date fixed by the notification.

(2) From the date so fixed such forest shall be deemed to be a reserved forest.

21. Publication of translation of such notification in neighbourhood of forest.–The Forest-officer shall, before the date fixed by such notification, cause a translation thereof into the local vernacular to be published in every town and village in the neighbourhood of the forest.

22. Power to revise arrangement made under section 15 or section 18–

The State Government may, within five years from the publication of any notification under section 20, revise any arrangement made under section 15 or section 18, and may for this purpose rescind or modify any order made under section 15 or section 18, and direct that any one of the proceedings specified in section 15 be taken in lieu of any other of such proceedings, or that the rights admitted under section 12 be commuted under section 16.

23. No right acquired over reserved forest, except as here provided.–

No right of any description shall be acquired in or over a reserved forest except by succession or under a grant or contract in writing made by or on behalf of the Government or some person in whom such right was vested when the notification under section 20 was issued.

24. Rights not to be alienated without sanction.–

(1) Notwithstanding anything contained in section 23, no right continued under clause (c) of sub-section (2) of section 15 shall be alienated by way of grant, sale, lease mortgage or otherwise, without the sanction of the State Government:

Provided that, when any such right is appendant to any land or house, it may be sold or otherwise alienated with such land or house.

(2) No timber or other forest-produce obtained in exercise of any such right shall be sold or bartered except to such extent as may have been admitted in the order recorded under section 14.

25. Power to stop ways and water-courses in reserved forests.–

The Forest-officer may, with the previous sanction of the State Government or of any officer duly authorised by it in this behalf, stop any public or private way or water-course in a reserved forest, provided that a substitute for the way or water-course so stopped, which the State Government deems to be reasonably convenient, already exists, or has been provided or constructed by the Forest-officer in lieu thereof.

26. Acts prohibited in such forests.–

(1) Any person who–

(a) makes any fresh clearing prohibited by section 5, or

(b) sets fire to a reserved forest, or, in contravention of any rules made by the State Government in this behalf, kindles any fire, or leaves any fire burning, in such manner as to endanger such a forest; or who, in a reserved forest–

(c) kindles, keeps or carries any fire except at such seasons as the Forest-officer may notify in this behalf,

(d) trespasses or pastures cattle, or permits cattle to trespass;

(e) causes any damage by negligence in felling any tree or cutting or dragging any timber;

(f) fells, girdles, lops, or bums any tree or strips off the bark or leaves from, or otherwise damages, the same;

(g) quarries stone, bums lime or charcoal, or collects, subjects to any manufacturing process, or removes, any forest-produce;

(h) clears or breaks up any land for cultivation or any other purpose;

(i) in contravention of any rules made in this behalf by the State Government hunts, shoots, fishes, poisons water or sets traps or snares; or

(j) in any area in which the Elephants' Preservation Act, 1879 (6 of 1879), is not in force, kills or catches elephants in contravention of any rules so made, shall be punishable with imprisonment for a term which may extend to six months, or with fine which may extend to five hundred rupees, or with both, in addition to such compensation for damage done to the forest as the convicting Court may direct to be paid.

(2) Nothing in this section shall be deemed to prohibit

(a) any act done by permission in writing of the Forest-officer, or under any rule made by the state Government; or

(b) the exercise of any right continued under clause (c) of sub-section (2) of section 15, or created by grant or contract in writing made by or on behalf of the Government under section 23.

(3) Whenever fire is caused wilfully or by gross negligence in a reserved forest, the State Government may (notwithstanding that any penalty has been inflicted under this section) direct that in such forest or any portion there of the exercise of all rights of pasture or to forest produce shall be suspended for such period as it thinks fit.

27. Power to declare forest no longer reserved.–(1) The State Government may, by notification in the Official Gazette, direct that, from a date fixed by such notification, any forest or any portion thereof reserved under the Act shall cease to be a reserved forest.

(2) From the date so fixed, such forest or portion shall cease to be reserved; but the rights (if any) which have been extinguished therein shall not revive in consequence of such cessation.

OF VILLAGE-FORESTS

28. Formation of village-forests.-

(1) The State Government may assign to any village-community the rights of Government to or over any land which has been constituted a reserved forest, and may cancel such assignment. All forests so assigned shall be called village-forests.

(2) The State Government may make rules for regulating the management of village forests, prescribing the conditions under which the community to which any such assignment is made may be provided with timber or other forest-produce or pasture, and their duties for the protection and improvement of such forest.

(3) All the provisions of this Act relating to reserved forests shall (so far as they are not inconsistent with the rules so made) apply to village-forests.

OF PROTECTED FORESTS

29. Protected forests.–

(1) The State Government may, by notification in the Official Gazette, declare the provisions of this Chapter applicable to

any forest-land or waste-land which,, is not included in a reserved forest but which is the property of Government, or over which the Government has proprietary rights, or to the whole or any part of the forest produce of which the Government is entitled.

(2) The forest-land and waste-lands comprised in any such notification shall be called a "protected forest".

(3) No such notification shall be made unless the nature and extent of the rights of Government and of private persons in or over the forest-land or waste-land comprised therein have been inquired into and recorded at a survey or settlement, or in such other manner as the State Government thinks sufficient. Every such record shall be presumed to be correct until the contrary is proved:

Provided that, if, in the case of any forest-land or waste land, the State Government thinks that such inquiry and record are necessary, but that they will occupy such length of time as in the meantime to endanger the rights of Government, the State Government may, pending such inquiry and record, declare such land to be a protected forest, but so as not to abridge or affect any existing rights of individuals or communities.

30. Power to issue notification reserving trees, etc.–

The State Government may, by notification in the Official Gazette,

(a) declare any trees or class of trees in a protected forest to be reserved from a date fixed by, the notification;

(b) declare that any portion of such forest specified in the notification shall be closed for such term, rot exceeding thirty years, as the State Government thinks fit, and that the rights of private persons, if any, over such portion shall be suspended during such terms, provided that the remainder of such forest be sufficient, and in a locality reasonably convenient, for the due exercise of the right suspended in the portion so closed; or

(c) prohibit, from a date fixed as aforesaid, the quarrying of stone, or the burning of lime or charcoal, or the collection or subjection to any manufacturing process, or removal of, any forest-produce in any such forest, and the breaking up or clearing for cultivation, for building, for herding cattle or for any other purpose, of any land in any such forest.

31. Publication of translation of such notification in neighbourhood.–

The Collector shall cause a translation into the local vernacular of every notification issued under section 30 to be affixed in a conspicuous place in every town and village in the neighbourhood of the forest comprised in the notification.

32. Power to make rules for protected forests.–The State Government may make rules to regulate the following matters, namely:

(a) the cutting, sawing, conversion and removal of trees and timber, and the collection, manufacture and removal of forest-produce, from protected forests;

(b) the granting of licences to the inhabitants of towns and villages in the vicinity of protected forests to take trees, timber or other forest-produce for their own use, and the production and return of such licences by such persons;

(c) the granting of licences to persons felling or removing trees or timber or other forest-produce from such forests for the purposes of trade, and the production

d) the payments, if any, to be made by the persons mentioned in clauses (b) and (c) for permission to cut such trees, or to collect and remove such timber or other forest-produce;

(e) the other payments, if any, to be made by them in respect of such trees, timber and produce, and the places where such payment shall be made;

(f) the examination of forest-produce passing out of such forests;

(g) the clearing and breaking up of land for cultivation or other purposes in such forests;

(h) the protection from fire of timber lying in such forests and of trees reserved under section 30;

(i) the cutting of grass and pasturing of cattle in such forests;

(j) hunting, shooting, fishing, poisoning water and setting traps or snares in such forests and the killing or catching of elephants in such forests in areas in which the Elephants' Preservation Act, 1879 (6 of 1879), is not in force;

(k) the protection and management of any portion of a forest closed under section 30; and

(l) the exercise of rights referred to in section 29.

33. Penalties for acts in contravention of notification under section 30 or of rules under section 32.—(1) Any person who commits any of the following offences, namely:–

(a) fells, girdles, lops, taps or bums any tree reserved under section 30, or strips off the bark or leaves from, or otherwise damages, any such tree;

(b) contrary to any prohibition under section 30, quarries any stone, or bums any lime or charcoal or collects, subjects to any manufacturing process, or removes any forest-produce;

(c) contrary to any prohibition under section 30, breaks up or clears for cultivation or any other purpose any land in any protected forest;

(d) sets fire to such forest, or kindles a fire without taking all reasonable precautions to prevent its spreading to any tree reserved under section 30, whether standing fallen or felled, or to say closed portion of such forest;

(e) leaves burning any fire kindled by him in the vicinity of any such tree or closed portion;

(f) fells any tree or drags any timber so as to damage any tree reserved as aforesaid;

(g) permits cattle to damage any such tree;

(h) infringes any rule made under section 32,

shall be punishable with imprisonment for a term which may extend to six months, or with fine which may extend to five hundred rupees, or with both.

(2) Whenever fire is caused wilfully or by gross negligence in a protected forest, the State Government may, notwithstanding that any penalty has been inflicted under this section, direct that in such forest or any portion thereof the exercise of any right of pasture or to forest-produce shall be suspended for such period as it thinks fit.

34. Nothing in this Chapter to prohibit acts done in certain cases.–

Nothing in this Chapter shall be deemed to prohibit any act done with the permission in writing of the Forest-officer, or in accordance with rules made under section 32, or, except as regards any portion of a forest closed under section 30, or as regards any rights the exercise of which has been suspended under section 33, in the exercise of any right recorded under section 29.

OF THE CONTROL OVER FORESTS AND LANDS NOT BEING THE PROPERTY OF GOVERNMENT

35. Protection of forests for special purposes.-(1) The State Government may, by notification in the Official Gazette, regulate or prohibit in any forest or waste-land

(a) the breaking up or clearing of land for cultivation;

(b) the pasturing of cattle; or

(c) the firing or clearing of the vegetation;

when such regulation or prohibition appears necessary for any of the following purposes:–

(i) for protection against storms, winds, rolling stones, floods and avalanches;

(ii) for the preservation of the soil on the ridges and slopes and in the valleys of hilly tracts, the prevention of land slips or of the formation of ravines, and torrents, or the protection of land against erosion, or the deposit thereon of sand, stones or gravel;

(iii) for the maintenance of a water-supply in springs, rivers and tanks;

(iv) for the protection of roads, bridges, railways and other lines of communication;

(v) for the preservation of the public health.

(2) The State Government may, for any such purpose, construct at its own expense, in or upon any forest or waste-land, such work as it thinks fit.

(3) No notification shall be made under sub-'section (1) nor shall any work be begun under sub-section (2), until after the issue of a notice to the owner of such forest or land calling on him to show cause, within a reasonable period to be specified in such notice, why such notification should not be made or work constructed, as the case may be, and until his objections, if any, and any evidence he may produce in support of the same, have been heard by an officer duly appointed in that behalf and have been considered by the State Government.

36. Power to assume management of forests.–

(1) In case of neglect of, or wilful disobedience to, any regulation or prohibition under section 35, or if the purposes of any work to be-constructed under that section so require, the State Government may, after notice in writing to the owner of such forest or land and after considering his objections, if any, place the same under the control of a Forest-officer, and may declare that all or any of the provisions of this Act relating to reserved forests shall apply to such forest or land.

(2) The net profits, if any, arising from the management of such forest or land shall be paid to the said owner.

37. Expropriation of forests in certain cases.–

(1) In any case under this Chapter in which the State Government considers that, in lieu of placing the forest or land under the control of a Forest-Officer, the same should be acquired for public purposes, the State 3overnment may proceed to acquire it in the manner provided by the Land Acquisition Act, 1894 (1 of 1894).

(2) The owner of any forest or land comprised in any notification under section 35 may, t any time not less than

three or more than twelve years from the date thereof, require that such forest or land shall be acquired for public purposes, and the State Government shall require such forest or land accordingly.

38. Protection of forests at request of owners.–

(1) The owner of any land or, if there more than one owner thereof, the owners of shares therein amounting in the aggregate at least two-thirds thereof may, with a view to the formation or conservation of forests thereon, represent in writing to the Collector their desire

(a) that such land be managed on their behalf by the Forest-officer as a reserved or a protected forest on such terms as may be mutually agreed upon; or

(b) that all or any of the provisions of this Act be applied to such land.

(2) In either case,-the State Government may, by notification in the Official Gazette, apply to such land such provisions of this Act as it thinks suitable to the circumstances thereof and as may be desired by the applicants.

OF THE DUTY ON TIMBER AND OTHER FOREST-PRODUCE

39. Power to impose duty on timber and other forest-produce.—(1) The 7[Central Government] may levy a duty in such manner, at such places and at such rates as it may declare by notification in the Official Gazette on all timber or other forest-produce

(a) which is produced in 8[the territories to which this Act extends], and in respect of which the Government has any right;

(b) which is brought from any place outside 8[the territories to which this Act extends.

(2) In every case in which such duty is directed to be levied ad valorem the 7[Central Government] may fix by like notification the value on which such duty shall be assessed.

(3) All duties on timber or other forest-produce which, at the time when this Act comes into force in any territory, are levied therein under the authority of the State Government, shall be deemed to be and to have been duty levied under the provisions of this Act.

10[(4) Notwithstanding anything in this section, the State Government may, until provision to the contrary is made by 11[Parliament], continue to levy any duty which it was lawfully levying before the commencement12 of 13[the Constitution], under this section as then in force:

Provided that nothing in this sub-section authorises the levy of any duty which as between timber or other forest-produce of the State and similar produce of the locality outside the State, discriminates in favour of the former, or which, in the case of timber or other forest-produce of localities outside the State, discriminates between timber or other forest-produce of one locality and similar timber or other forest-produce of another locality.]

40. Limit not to apply to purchase-money or royalty.-

Nothing in this Chapter shall be deemed to limit the amount, if any, chargeable as purchase-money or royalty on any timber or other forest-produce, although the same is levied on such timber or produce while in transit, in the same manner as duty is levied.

OF THE CONTROL OF TIMBER AND OTHER FOREST-PRODUCE IN TRANSIT

41. Power to make rules to regulate transit of forest produce.—

(1) The control of all rivers and their banks as regards the floating of timber, as well as the control of all timber and other forest-produce in transit by land or water, is vested in the State Government, and it may make rules to regulate the transit of all timber and other forest-produce.

(2) In particular and without prejudice to the generality of the foregoing power such rules may

(a) prescribe the routes by which alone timber or other forest-produce may be imported, exported or moved into, from or within 14[the State];

(b) prohibit the import or export or moving of such timber or other produce without a pass from an officer duly authorised to issue the same, or otherwise than in accordance with the conditions of such pass;

(c) provide for the issue, production and return of such passes and for the payment of fees therefor;

(d) provide for the stoppage, reporting, examination and marking of timber or other forest-produce in transit, in respect of which there is reason to believe that any money is payable to the Government on account of the price thereof, or on account of any duty, fee, royalty or charge due thereon, or, to which it is desirable for the purposes of this Act to affix a mark;

(e) provide for the establishment and regulation of depots to which such timber or other produce shall be taken by those in charge of it for examination, or for the payment of such money, or in order that such marks may be affixed to it, and the conditions under which such timber or other produce shall be brought to, stored at and removed from such depots;

(f) prohibit the closing up or obstructing of the channel or banks of any river used for the transit of timber or other forest-produce, and the throwing of grass, brushwood, branches or leaves into any such river or any act which may cause such river to be closed or obstructed;

(g) provide for the prevention or removal of any obstruction of the channel or banks of any such river, and for recovering the cost of such prevention or removal from the person whose acts or negligence necessitated the same;

(h) prohibit absolutely or subject to conditions, within specified local limits, the establishment of sawpits, the converting, cutting, burning, concealing or making of timber, the altering or effacing of any marks on the same, or the possession or carrying of marking hammers or other implements used for marking timber;

(i) regulate the use of property marks for timber, and the registration of such marks; prescribe the time for which such r6gistration'shall hold good; limit the number of such marks that may be registered by any one person, and provide for the levy of fees for such registration.

(3) The State Government may direct that any rule made under this section shi-11 not apply to any specified class of timber or other forest-produce or to any specified local area.

[41A. Powers of Central Government as to movements of timber across customs frontiers.—

Notwithstanding anything in section 41, the Central Government may make rules to prescribe the route by which alone timber or other forest-produce may be imported, exported or moved into or from 16[the territories to which this Act extends) across any customs frontier as defined by the Central Government, and any rules made under section 41 shall have effect subject to the rules made under this section.

42. Penalty for breach of rules made under section 41.-

(1) The State Government may by such rules prescribe as penalties for the contravention thereof imprisonment for a term which may extend to six months, or fine which may extend to five hundred rupees, or both.

(2) Such rules may provide that penalties which are double of those mentioned in subsection (1) may be inflicted in cases where the offence is committed after sunset and before sunrise, or after preparation for resistance to lawful authority, or where the offender has been previously convicted of a like offence.

43. Government and Forest-officers not liable for damage to forest-produce at depot.-

The Government shall not be responsible for any loss or damage which may occur in respect of any timber or other forest-produce while at a depot established under a rule made under section 41, or while detained elsewhere, for the purposes of this Act; and no Forest-officer shall be responsible for any such loss or damage, unless he causes such loss or damage negligently, maliciously or fraudulently.

44. All persons bound to aid in case of accidents at depot.–

In case of any accident or emergency involving danger to any property at any such depot, every person employed at such depot, whether by the Government or by any private person, shall render assistance to any Forest-officer or Police-officer demanding his aid in averting such danger or securing such property from damage or loss.

OF THE COLLECTION OF DRIFT AND STRANDED TIMBER

45. Certain kinds of timber to be deemed property of Government until title thereto proved, and may be collected accordingly.—

(1) All timber found adrift, beached, stranded or sunk; all wood or timber bearing marks which have not been registered in accordance with the rules made under section 41, or on which the marks have been obliterated, altered or defaced by fire or otherwise; and in such areas as the State Government directs, all unmarked wood and timber, shall be deemed to be the property of Government, unless and until any person establishes his right and title thereto, as provided in this Chapter.

(2) Such timber may be collected by any Forest-officer or other person entitled to collect the same by virtue of any rule made under section 51 and may be brought to any depot which the Forest-officer may notify as a depot for the reception of drift timber.

(3) The State Government may, by notification in the Official Gazette, 6xempt any class of timber from the provisions of this section.

46. Notice to claimants of drift timber.-

Public notice shall from time to time be given by the Forest-officer, of timber collected under section 45. Such notice shall contain a description of the timber, and shall require any person claiming the same to present to such officer, within a period not less than two months from the date of such notice, a written statement of such claim.

47. Procedure on claim preferred to such timber.-

(1) When any such-statement is presented as aforesaid, the Forest-officer may, after making such inquiry as he thinks fit, either reject the claim after according his reasons for so doing, or deliver the timber to the claimant.

(2) If such timber is claimed by more than one person, the Forest-officer may either deliver the same to any of such persons who he deems entitled thereto, or may refer the claimants to the Civil Courts, and retain the timber pending the receipt of an order from any such Court for its disposal.

(3) Any person whose claim has been rejected under this section may, within three months from the date of such rejection, institute a suit to recover possession of the timber claimed by him; but no person shall recover any compensation or costs against the Government, or against any Forest-officer on account of such rejection, or the detention or removal of any timber, or the delivery thereof to any other person under this section.

(4) No such timber shall be subject to process of any Civil, Criminal or Revenue Court until it has been delivered, or a suit has been brought, as provided in this section.

48. Disposal of unclaimed timber.-

If no such statement is presented as aforesaid, if the claimant omits to prefer his claim in the manner and within the period fixed by the notice issued under section 46, or on such claim having been so preferred by him and having been rejected, omits to institute a suit to recover possession of such timber within the further period fixed by section 47, the ownership of such timber shall vest in the Government, or, when such timber has been delivered to another person under section 47, in such other person free from all encumbrances not created by him.

49. Government and its officers not liable for damage to such timber.-

TN Government shall not be responsible for any loss or damage which may occur in respect any timber collected under section 45, and no Forest-officer shall be responsible for any such loss or damage, unless he causes such loss or damage negligently, maliciously fraudulently,

50. Payments to be made by claimant before timber is delivered to him.-

No person shall be entitled to recover possession of any timber collected or delivered as aforesaid until he has paid to the Forest-officer or other person entitled to receive it such sum on account thereof as may be due under any rule made under section 51.

51. Power to make rules and prescribe penalties.-

(1) The State Government in make rules to regulate the following matters, namely:

(a) the salving, collection and disposal of all timber mentioned in section 45;

(b) the use and registration of boats used in salving and collecting timber;

(c) the amounts to be paid for salving, collecting, moving, storing or disposing such timber; and

(d) the use and registration of hammers and other instruments to be used marking such timber.

(2) The State Government may prescribe, as penalties for the contravention of any rule made under this section, imprisonment for a term which may extend to six months, or fine which may extend to five hundred rupees, or both.

PENALTIES AND PROCEDURE

52. Seizure of property liable to confiscation.-(1) When there is reason to believe that a forest-offence has been committed in respect of any forest-produce, such produce together with all tools, boats, carts or cattle used in committing any such offence, may be seized by any Forest-officer or Police-officer.

(2) Every officer seizing any property under this section shall place on such property a mark indicating that the same has been so seized, and shall, as soon as may be, make a - report

of such seizure to the Magistrate having jurisdiction to try the offence on account which the seizure has been made:

Provided that, when the forest-produce with respect to which such offence is belie to have been committed is the property of Government, and the offender is unknown, it shall be sufficient if the officer makes, as soon as may be, a report of the circumstances to his official superior.

53. Power to release property seized under section 52.-Any Forest-officer of a rank not inferior to that of a Ranger who, or whose subordinate, has seized any tools, boats, carts or cattle under section 52, may release the same on the execution by the owner thereof a bond for the production of the property so released, if and when so required, before the Magistrate having jurisdiction to try the offence on account of which the seizure has been made.

54. Procedure thereupon.-Upon the receipt of any such report, the Magistrate shall, with all convenient despatch, take such measures as may be necessary for the arrest and trial of the offender and the disposal of the property according to law.

55. Forest-produce, tools, etc., when liable to confiscation.— (1) All timber or forest produce which is not the property of Government and in respect of which a forest-offence has been committed, and all tools, boats, carts and cattle used in committing any forest offence, shall be liable to confiscation.

(2) Such confiscation may be in addition to any other punishment prescribed for such offence.

56. Disposal on conclusion of trial for forest-offence, of produce in respect of which it was committed.-When the trial of any forest-offence is concluded, any forest-produce in respect of which such offence has been committed shall, if it is the property of Government or has been confiscated, be taken charge of by a Forest-officer, and, in any other case, may be disposed of in such manner as the Court may direct.

57. Procedure when offender not known or cannot be found.-When the offender is not known or cannot be found, the Magistrate may, if he finds that an offence has been committed,

order the property in respect of which the offence has been committed to be confiscated and taken charge of by the Forest-officer, or to be made over to the person whom the Magistrate deems to be entitled to the same:

Provided that no such order shall be made until the expiration of one month from d date of seizing such property, or without hearing the person, if any, claiming any rig thereto, and the evidence, if any, which he may produce in support of his claim.

58. Procedure as to perishable property siezed under section 52.-The Magistrate may, notwithstanding anything hereinbefore contained, direct the sale of any property seized under section 52 and subject to speedy and natural decay, and may deal with the proceeds as he would have dealt with such property if it had not been sold.

59. Appeal from orders under section 55, section 56 or section 57.-The officer made the seizure under section 52, or any of his official superiors, or any person claim to be interested in the property so seized, may, within one month from the date of any order passed under section 55, section 56 or section 57, appeal therefrom to the Court to will orders made by such Magistrate are ordinarily appealable, and the order passed on A appeal shall be final.

60. Property when to vest in Government.-When an order for the confiscation any property has been passed under section 55 or section 57, as the case may be, and the period limited by section 59 for an appeal from such order has elapsed, and no such an apppeal has been preferred, or when, on such an appeal being preferred, the Appellate C4 confirms such order in respect of the whole or a portion of such property, such property or such portion thereof, as the case may be, shall vest in the Government free from all incumbrances.

61. Saving of power to release property seized.-Nothing hereinbefore contained shall be deemed to prevent any officer empowered in this behalf by the State Government, from directing at any time the immediate release of any property seized under section 52.

62. Punishment for wrongful seizure.-Any Forest-officer or Police-officer who vexatiously and unnecessarily seizes any property on pretence of seizing property liable confiscation under this Act shall be punishable with imprisonment for a term which extend to six months, or with fine which may extend to five hundred rupees, or with both.

63. Penalty for counterfeiting or defacing marks on trees and timber and for altering boundary marks.-Whoever, with intent to cause damage or injury to the public or to any person, or to cause wrongful gain as defined in the Indian Penal Code–

(a) knowingly counterfeits upon any timber or standing tree a mark used by Forest-officers to indicate that such timber or tree is the property o Government or of some person, or that it may lawfully be cut or removed some person; or

(b) alters, defaces or obliterates any such mark placed on a tree or on timber by under the authority of a Forest-officer; or

(c) alters, moves, destroys or defaces any boundary-mark of any forest or waste land to which the provisions of this Act are applied,

shall be punishable with imprisonment for a term which may extend to two years, or fine, or with both.

64. Power to arrest without warrant.—(1) Any Forest-officer or Police-officer without orders from a Magistrate and without a warrant, arrest any person against whom a reasonable suspicion exists of his having been concerned in any forest-offence punishable with imprisonment for one month or upwards.

(2) Every officer making an arrest under this section shall, without unnecessary delay and subject to the provisions of this Act as to release on bond, take or send the person arrested before the Magistrate having jurisdiction in the case, or to the officer in charge of the nearest police station.

(3) Nothing in this section shall be deemed to authorise such arrest for any act which is an offence under Chapter IV

unless such act has been prohibited under clause (c) of section 30.

65. Power to release on a bond a person arrested.-Any Forest-officer of a rank not inferior to that of a Ranger, who, or whose subordinate, has arrested any person under the provisions of section 64, may release such person on his executing a bond to appear, if and when so required, before the Magistrate having jurisdiction in the case, or before the officer in charge of the nearest police station.

66. Power to prevent commission of offence.-Every Forest-officer and Police officer shall prevent, and may interfere for the purpose of preventing, the commission of any forest-offence.

67. Power to try offences summarily.-The District Magistrate or any Magistrate of the first class specially empowered in this behalf by the State Government may try summarily, under the Code of Criminal Procedure, 1898, any forest-offence punishable with imprisonment for a term not exceeding six months, or fine not exceeding five hundred rupees, or both.

68. Power to compound offences.-(1) The State Government may, by notification in the Official Gazette, empower a Forest officer–

(a) to accept from any person against whom a reasonable suspicion exists that he has committed any forest-offence, other than an offence specified in section 62 or section 63, a sum of money by way of compensation for the offence which such person is suspected to have committed, and

(b) when any property has been seized as liable to confiscation, to release the same on payment of the value thereof as estimated by such officer.

(2) On the payment of such sum of money, or such value, or both, as the case may be, to such officer, the suspected person, if in custody, shall be discharged, the property, if any seized shall be released, and no further proceedings shall be taken against such person or property.

(3) A Forest-officer shall not be empowered under this section unless he is a Forest officer of a rank not inferior to

that of a Ranger and is in receipt of a monthly salary amounting to at least one hundred rupees, and the sum of money accepted as compensation under clause (a) of sub-section (1) shall in no case exceed the sum of fifty rupees.

69. Presumption that forest-produce belongs to Government.-When in any proceedings taken under this Act, or in consequence of anything done under this Act, a question arises as to whether any forest-produce is the property of the Government, such produce shall be presumed to be the property of the Government until the contrary is proved.

CATTLE-TRESPASS

70. Cattle-trespass Act, 1871, to apply.-Cattle trespassing in a reserved forest any portion of a protected forest which has been lawfully closed to grazing shall be deem to be cattle doing damages to a public plantation within the meaning of section II of Cattle-trespass Act, 1871 (1 of 1871), and may be seized and impounded as such by Forest-officer or Police-officer.

71. Power to alter fines fixed under that Act.-The State Government may, notification in the Official Gazette, direct that, in lieu of the fines fixed under section I the Cattle-trespass Act, 1871 (1 of 1871), there shall be levied for each head of cattle impounded under section 70 of this Act such fines as it thinks fit, but not exceeding following, that is to say:-

For each elephant ten rupees

For each buffalo or camel two rupees

For each horse, mare, gelding, pony, colt, filly, mule, bull,

bullock, cow, or heifer one rupee

For each calf, ass, pig, ram, ewe, sheep, lamb, goat or kid eight annas

OF FOREST-OFFICERS

72. State Government may invest Forest-officers with certain powers.-

(1) The State Government may invest any Forest-officer with all or of the following powers, that is to say:-

(a) power to enter upon any land and to survey, demarcate and make a map of the same;

(b) the powers of a Civil Court to compel the attendance of witnesses and production of documents and material objects;

(c) power to issue a search-warrant under the Code of Criminal Procedure, 1898 (5 of 1898); and

(d) power to hold an inquiry into forest-offences, and, in the course of such inquiry, to receive and record evidence.

(2) Any evidence recorded under clause (d) of sub-section (1) shall be admissible in subsequent trial before a Magistrate, provided that it has been taken in the presence of accused person.

73. Forest officers deemed public servants.-All Forest-officers shall be deemed be public servants within the meaning of the Indian Penal Code, 1860 (45 of 1860).

74. Indemnity for acts done in good faith.-No suit shall lie against any public servant for anything done by him in good faith under this Act.

75. Forest-officers not to trade.-Except with the permission in writing of the State Government, no Forest-officer shall, as principal or agent, trade in timber or other forest produce, or be or become interested in any lease of any forest or in any contract for working any forest, whether in or outside 17[the territories to which this Act extends].

SUBSIDIARY RULES

76. Additional powers to make rules.-

The State Government may make rules

(a) to prescribe and limit the powers and duties of any Forest-officer under this Act;

(b) to regulate the rewards to be paid to officers and informers out of the proceeds of fines and confiscation under this Act;

(c) for the preservation, reproduction and disposal of trees and timber belonging to Government, but grown on lands belonging to or in the occupation of private persons; and

(d) generally, to carry out the provisions of this Act.

77. Penalties for breach of rules.-Any person contravening any rule under this Act, for the contravention of which no special penalty is provided, shall be punishable with imprisonment for a term which may extend to one month, or fine which may extend to five hundred rupees, or both.

78. Rules when to have force of law.-All rules made by the State Government under this Act shall be published in the Official Gazette, and shall thereupon, so far as they are consistent with this Act, have effect as if enacted therein.

MISCELLANEOUS

79. Persons bound to assist Forest-officers and Police-officers.-

(1) Every person who exercises any right in a reserved or protected forest, or who is permitted to take any forest-produce from, or to cut and remove timber or to pasture cattle in, such forest, and every person who is employed by any such person in such forest, and every person in any village contiguous to such forest who is employed by the Government or who receives emoluments from the Government for services to be performed to the community, shall be bound to furnish without unnecessary delay to the nearest Forest-officer or Police officer any information he may possess respecting the commission of, or intention to commit, any forest-offence, and shall forthwith take steps, whether so required by any Forest-officer or Police officer or not,-

(a) to extinguish any forest fire in such forest of which he has knowledge or information;

(b) to prevent by any lawful means in his power any fire in the vicinity of such forest of which he has knowledge or information from spreading to such forest, and shall assist any Forest-officer or Police-officer demanding his aid–

(c) in preventing the commission in such forest of any forest-offence; and

(d) when there is reason to believe that any such offence has been committed such forest in discovering and arresting the offender.

(2) Any person who, being bound so to do, without lawful excuse (the burden ofprovin which shall lie upon such person) fails–

(a) to furnish without unnecessary delay to the nearest Forest-officer or Polic(officer any information required by sub-section (1);

(b) to take steps, as required by sub-section (1), to extinguish any forest fire in reserved or protected forest;

(c) to prevent, as required by sub-section (1), any fire in the vicinity of such fore.4 from spreading to such forest or

(d) to assist any Forest-officer or Police officer demanding his aid in preventing th commission in such forest of any forest-offence, or, when there is reason believe that any such offence has been committed in such forest, in discover' and arresting the offender, shall be punishable with imprisonment for a term which extend to one month, or with fine which may extend to two hundred rupees, or with both.

80. Management of forests the joint property of Government and other persons.–(1) If the Government and any person be jointly interested in any forest or waste-land, a in the whole or any part of the produce thereof, the State Government may either–

(a) undertake the management of such forest, waste-land or produce, accounting to such person for his interest in the same; or

(b) issue such regulations for the management of the forest, waste-land or produce by the person so jointly interested as it deems necessary for the management thereof and the interests of all parties therein.

(2) When the State Government undertakes under clause (a) of sub-section (1) the management of any forest, waste-land

or produce, it may, by notification in the Official Gazette, declare that any of the provisions contained in Chapters 11 and IV shall apply to such forest, waste-land or produce, and thereupon such provisions shall apply accordingly.

81. Failure to perform service for which a share in produce of Government forest is employed.-If any person be entitled to a share in the produce of any forest which is the property of Government or over which the Government has proprietary rights or to any part of the forest-produce of which the Government is entitled upon the condition of duly performing any service connected with such forest, such share shall be liable to confiscation in the event of the fact being established to the satisfaction of the State Government that such service is no longer so performed:

Provided that no such share be confiscated until the person entitled thereto, and the evidence, if any, which he may produce in proof of the due performance of such service, have been heard by an officer duly appointed in that behalf by the State Government.

82. Recovery of money due to Government.—All money payable to the Government under this Act, or under any rule made under this Act, or on account of the price of any forest produce, or of expenses incurred in the execution of this Act in respect of such produce, may, if not paid when due, be recovered under the law for the time being in force as if it were an arrear of land-revenue.

83. Lien on forest-produce for such money.—(1) When any such money is payable for or in respect of any forest-produce, the amount thereof shall deemed to be a first charge on such produce, and such produce may be taken possession of by a Forest-officer until such amount has been paid.

(2) If such amount is not paid when due, the Forest-officer may sell such produce by public auction, and the proceeds of the sale shall be applied first in discharging such amount.

(3) The surplus, if any, if not claimed within two months from the date of the sale by the person entitled thereto, shall be forfeited to Government.

84. Land required under this Act to be deemed to be needed for a public purpose under the Land Acquisition Act, 1894.—

Whenever it appears to the State Government that any land is required for any of the purposes of this Act, such land shall be deemed to be needed for a public purpose within the meaning of section 4 of the Land Acquisition Act, 1894 (1 of 1894).

85. Recovery of penalties due under bond.–When any person, in accordance with any provision of this Act, or in compliance with any rule made thereunder, binds himself by any an bond or instrument to perform any duty or act or covenants by any bond or instrument that he, or that he and his servants and agents will abstain from any act, the whole sum mentioned in such bond or instrument as the amount to be paid in case of a breach of the conditions thereof may, notwithstanding anything in section 74 of the Indian Contract Act, 1872 (9 of 1872), be recovered from him in case of such breach as if it were an arrear of land revenue.

85A. Saving for rights of Central Government.-Nothing in this Act shall authorise a Government of any State to make any order or do anything in relation to any property not vested in that State or otherwise prejudice any rights of the Central Government or the Government of any other State without the consent of the Government concerned.

86. Repeals.- [Rep. by Repealing and Amending Act, 1947 (2 of 1948), sec.2 and Sch.]

NAME OF THE STATE/UNION TERRITORY ACT

A Bill to endow the ownership rights of Minor Forest Produce to the Forest Dependent Communities.

Preamble – WHEREAS around Twenty-Eight percent population of the country, including Scheduled Tribe population, living in about Sixty-One Thousand villages in and around Forests depend for their subsistence and livelihood on forests;

AND WHEREAS there is a large population of tribal and non tribal landless and economically backward communities in the state who derive their livelihood security from the forests of the state;

AND WHEREAS the minor forest produce is the main source of livelihood of communities dependent on forests and there is need for the sustainable utilization and long-term conservation of the minor forest produce found in the forests of the state;

AND WHEREAS it has been recognised at the national level that the collection, processing and trade in minor forest produce should be endowed with the forest dependent communities on the principle of 'share and care'

BE it enacted by the State/ UT legislature in the Fifty-fifth year of the Republic of India as follows: -

1.Short title, extent, and commencement:

(1) This Act may be called the State/Union Territory Minor Forest Produce (Ownership of Forest Dependent Community) Act, 2005.

(2) It extends to the Government forest lands in the whole of the State/Union Territory, subject to the provisions under the Wildlife (Protection) Act, 1972, Biological Diversity Act, 2002 and other State Forest Acts, as notified by the State/Union Territory Government,

(3) It shall come into force on such date or dates as the State/Union Territory Government may, by notification, specify in this behalf.

Explanation: Under sub section (2) . State Government may exclude National Parks and Sanctuaries because of Supreme Court order banning commercial harvesting of forest produce and other lands depending on state specific situation which may vary from state to state.

2.Definitions:-

In this Act, unless the context otherwise requires,

(a) "Forest Dependent Community" means a group of individuals constituted into a Joint Forest Management Committee, and / or any other group of individuals, including Gram Sabha, Village council, Eco Development Committee, dependent on forests notified by the State/Union Territory Government as "Forest Dependent Community" for the purposes of this Act.

(b) "Forest Offence" means the forest offence as defined in clause (3) of Section 2 of the Indian Forest Act, 1927 (No. XVI of 1927) (or concerned State Forest Act).

(c) "Joint Forest Management Committee" means a committee constituted as such in accordance with the provisions of the Joint Forest Management resolution of the State/Union Territory, and recognised by the State/Union Territory Forest Department. The words Joint Forest Management Committee shall be synonymous with terms - Van Samrakshan Samiti, Village Forest Committee, Forest Protection Committee or any other such committee as notified by the State/Union Territory Government.

(d) "Minor Forest Produce" means all non-timber forest produce of plant origin as notified by the State/Union Territory as Minor Forest Produce, and will include honey, lac, wax and tusser, in the official gazette, from time to time, for the purposes of this Act.

(e) "Ownership" means right to collect and process the Minor Forest Produce in a manner prescribed by the State/ Union Territory Government and its consumption, or sale to others, by the holder of the right.

(f) "Right Holder" means a person who has been given any right over forest produce at the time of forest settlement proceeding as published in the Government Gazette.

(g) "Working Plan" is a silvicultural management plan of the forest division, prepared on the basis of Working Plan Code, prescribed by the Central Government.

(h) "Sustained yield" is the quantity of minor forest produce , which can be harvested on a sustained basis from any forest land (s) without impairing the long-term productivity.

(i) "Non-destructive harvesting" means the harvesting methodology and technique on the basis of sustained yield principle, as prescribed in the working plan, that does not cause irreversible damage to the plants and the forest.

(j) "Carrying capacity" means the maximum quantity of the minor forest produce that can be harvested from any forest land (s).

(k) "Net income" means the income derived after deducting all expenses on collection, trade, value addition and any other cost from the total income from the trade in minor forest produce, in a given area.

(l) "Authorised officer" means an officer as notified by the State/ UT Government for the purpose of this Act.

Explanation: Under clause (a) Forest Dependent Community has been defined as Joint Forest Management Committee because under State JFM resolutions these have been given the responsibility of protection, development and benefit sharing. State Governments have been empowered to identify other groups also as Forest Dependent Community such as Forest Panchayat, Gram Sabha, Village Council, Eco-development committee and cooperative societies etc. Under clause (e) the minor forest produce will not include forest produce of animal origin except lac, wax, honey, tusser because of the provisions of Wildlife (Protection) Act. Bamboos and canes have also been excluded till an amendment is brought in the Indian Forest Act to exclude these from the category "Tree". However State/Union Territory Government are encouraged to give ownership rights to Bamboos & Canes also in areas adjoining the villages, through an executive order, till an amendment in Indian Forest Act, 1927 is brought about.

3. Declaration of Forest Dependent Community

The State/Union Territory Government shall notify in the official gazette, from time to time, different categories of group of individuals dependent on forests as a "Forest Dependent Community" for the purposes of this Act.

Explanation: This has been left to the State/Union Territory Government because of State/Union Territory specific situation, which may vary, from State/Union Territory to State/Union Territory as explained under sectio 3 above.

4.Declaration of Minor Forest Produce

(a)The State/Union Territory Government shall notify in the official gazette, a list of "Minor Forest Produce" which shall be considered as "Minor Forest Produce" for the purposes of this Act.

(b) The State/Union Territory Government may, from time to time, by notification, direct that from a date specified therein, a Minor forest produce specified therein, shall cease to be a minor forest produce in relation to the area or areas as may be specified in the said notification

(c) The State/Union Territory Government may, from time to time, by a like notification, direct that from a date specified therein the specified minor forest produce which ceased to be so under sub section (a) above, shall be minor forest produce in relation to the area or areas as may be specified in the said notification

Explanation: This has been left to the State/Union Territory Government because of State/Union Territory specific situation which may vary from State/Union Territory to State/ Union Territory . The reason could be the status of any minor forest produce at a given point of time such as rare, endangered, critically endangered etc.

5. Endowment of ownership of minor forest produce through assignment of areas:

The Forest Dependent Communities are endowed with the ownership of minor forest produce found on Government forest land(s) for which, the State/Union Territory Government shall, by notification, entrust a specified forest area to a Forest Dependent Community, and that different Forest Dependent Communities are entrusted different areas.

Explanation: State/Union Territory Governments have been empowered to assign different areas to different Forest Dependent Communities to avoid any confusion, duplication and also in consideration of other possible practical difficulties which may be State/Union Territory specific.

6. Collection of Minor Forest Produce

Once endowed with ownership of minor forest produce, the Forest Dependent Community may collect the specified minor forest produces from the area entrusted to it in a manner prescribed by the State/Union Territory Government.

Provided that the harvesting of minor forest produce shall be on a non-destructive and sustainable basis, based on the carrying capacity of the area and provisions of the working plans.

Provided further that meeting the needs of domestic requirements of Forest Dependent Communities/Right Holders of the area shall be the first charge on the minor forest produce of the forest area entrusted to the Forest Dependent Community.

Explanation: *The section provides for undertaking collection, extraction and harvesting on a sustainable and non destructive basis. It also provides for safeguarding the interests of meeting the domestic requirement and the privileges of the right holders.*

7. Trade in Minor Forest Produce:

(a) The trade in respect of such minor forest produce, for which any law has been enacted by the State/Union Territory Government, will be in accordance with the provisions of the said law, so long as the said law is in force.

(b) The trade in respect of such minor forest produce for which no specific law has been made, shall be in accordance with the procedure, as prescribed by the Forest Dependant Community.

Explanation: Though the intention is to endow ownership of all minor forest produce to the Forest Dependent Community but taking into account the fact that in many states the trade in certain minor forest produce has been nationalised and it may not be possible to switch over to new system at once, the provision has been made to continue with the existing system till new arrangement is made and issues related to revenue to the state and the fate of the staff engaged in the management of existing system are properly addressed. The efforts should begin in the direction of repealing the existing laws and extend the provisions of this law to all minor forest produce as early as possible.

8. Sharing of net income from the trade in Minor Forest Produce

The net income from the trade in minor forest produce shall be deposited in the account of the Forest Dependent Community and shall be utilised, for the purposes mentioned as under, on the basis of an action plan approved by the Forest Dependent Community.

(a) Not less than twenty five per cent on regeneration and development of forest area entrusted.

(b) Not more than seventy five percent for distribution among the members of the Forest Dependent Community as may be decided by the Forest Dependant Community. Provided that the Forest Dependent Community shall have the power to reappropriate any amount of income available under clause (b) for the purposes of clause (a) .

Explanation: Provision has been made for the conservation and development of assigned forest also.

9. Duties and responsibilities of the Forest Dependent Community

(a) It shall be the duty of the Forest Dependant Community, to identify themselves with the protection and development of the forest areas specified under this Act in the manner notified by the State/Union Territory Government

(b) The State Govt may by notification authorise Forest Dependent Community to take cognisance of a forest offence committed by any person in the forest area entrusted to the Forest Dependent Community.

Explanation: Provision has been made to ensure the protection of forests assigned to the Forest Dependant Community.

10. Other Matters

The matters related to administration, planning, accounts, audit, inspections, committees and functioning of the Forest Dependent Community shall be dealt in a manner prescribed by the State/Union Territory Government.

Explanation: This section has been provided to empower the Government to prescribe procedures for the guidance of the Forest dependent Community

11. Attempts and abetment:

Any person who attempts to contravene or abets the contravention of any of provisions of this Act or the rules made there under shall be deemed to have contravened such provisions.

Explanation: This is a standard clause.

12. Protection of action taken in good faith: -

(a) No suit, prosecution or other legal proceedings shall lie against any person for anything which is done in good faith or intended to be so done in pursuance of this Act or the rules made there under.

(b) No suit, prosecution or other legal proceedings shall lie against the State Government for any damage caused or likely to be caused or any injury suffered or likely to be suffered by virtue of the provisions of this Act, or by anything which is done in good faith or intended to be so done in pursuance of this Act or the rules made there under.

Explanation: This is a standard clause.

13. Power to make rules: -

(1) The State/Union Territory Government may, by notification, make rules for carrying out all or any of the provisions of this Act.

(2) In particular and without prejudice to the generality of the foregoing provisions such rules may provide for all or any of the following matters, namely;

(a) Identification and listing of group of individuals who will be considered to be the Forest Dependent Community

(b) Identification and listing of the forest produce which will be considered as Minor forest Produce

(c) Endowment of ownership of Minor Forest Produce and assignment of areas of the Forest Dependent Community

(d) Sustainable collection, extraction and transport of Minor forest Produce by Forest Dependent Community from the areas assigned to it

(e) Trade in respect of such Minor Forest Produce whose trade is not nationalised

(f) Duties and responsibilities of Forest Dependent Community

(g) Powers and functions of the Forest Dependent Community

(h) Administration, Planning, Budget, Accounts, Audit, Inspections, and Committees related to the functioning of the Forest Dependent Community.

(i) Any other matter which is either expressly or implied required to be prescribed under this Act.

(3) Every rule made under this Act shall be laid as soon as may be after it is made before each House of the State Legislature, while it is under session.

Explanation: It has been left to State/Union Territory Governments to make rules for various matters as the situation may vary from state to state

14. (a) Violation under the Act: Any violation under this act, shall be dealt in accordance with the provisions under Indian Forest Act, 1927 and/or concerned State Forest Act (s).

Explanation: This section has been provided to deal with the forest offences.

(b) Notwithstanding anything contained in any other Act to the contrary, the authorized officer will have the power to suitably restrain the activities of any forest dependent community if the provisions of this Act are wilfully violated by the forest dependent community to the detriment and irreparable damage to the natural resource base of the entrusted area .

Explantion: This provision is added to ensure that the communities enjoying the rights do not undermine the resource base under the influence of forces.

15.Removal of difficulties:-

If any difficulty arises in giving effect, to the provisions of this Act the State Government may, by order, do anything not inconsistent with such provisions, which appear to it to be necessary or expedient for the purposes of removing the difficulty.

Explanation: This is a standard clause.

16.Saving:

It is hereby declared that any thing done or any action taken (including any notification, order, notice etc.) under the provisions of existing Acts relating to minor forest produce, be deemed to have been done or taken under the corresponding provisions of this Act as if this Act were in force at the time of such thing was done or action was taken and shall continue to be in force, unless and until superseded by any thing done or any action taken under this Act.

Explanation: This is a standard clause.

8

Ministry of Environment and Forests

NOTIFICATION

New Delhi, the 10th January, 2003

G.S.R.23(E):- In exercise of the powers conferred by sub-section (1) of section 4 of the Forest (Conservation) Act, 1980 (69 of 1980), and in supersession of the Forest (Conservation) Rules, 1981, except as respects things done or omitted to be done before such supersession, the Central Government hereby makes the following rules, namely:-

1. Short title, extent and commencement.-

1. These rules may be called the Forest (Conservation) Rules, 2003.

2. They shall extend to the whole of India except the State of Jammu and Kashmir.

3. They shall come into force on the date of their publication in the Official Gazette.

2. Definitions.- In these rules, unless the context otherwise requires:-

a. "Act" means the Forest (Conservation) Act, 1980 (69 of 1980);

b. "Committee" means the Advisory Committee constituted under section 3 of the Act;

c. "Chairperson" means the Chairperson of the Committee;

d. "Member" means a member of the Committee;

e. "Nodal Officer" means any officer not below the rank of Conservator of Forests, authorised by the State Government to deal with the forest conservation matters under the Act;

f. "Regional Office" means a Regional Office of the Central Government in the Ministry of Environment and Forests established as part of the Ministry to deal with the forest conservation matters under the Act;

g. "Section" means a section of the Act;

h. "User Agency" means any person, organisation or Company or Department of the Central or State Government making a request for diversion or de-notification of forest land for non-forest purpose or using forest land for non-forest purpose in accordance with the permission granted by the Central Government under the Act or the rules.

3. Composition of the Committee.-

(1) The Committee shall be composed of the following members:-

i. Director General of Forests, Ministry of Environment and Forests - Chairperson.

ii. Additional Director General of Forests, Ministry of Environment and Forests- Member.

iii. Additional Commissioner (Soil Conservation), Ministry of Agriculture- Member.

iv. Three eminent experts in forestry and allied disciplines (non-officials)- Members.

v. Inspector General of Forests (Forest Conservation), Ministry of Environment and Forests - Member Secretary

(2) Additional Director General of Forests shall act as the Chairperson in the absence of Director General of Forests.

4. Terms of appointment of non-official members shall be as follows.-

i. a non-official member shall hold his office for a period of two years;

ii. a non-official member shall cease to hold office if he becomes of unsound mind, becomes insolvent or is convicted by court of law on a criminal offence involving moral turpitude;

iii. a non-official member may be removed from his office if he fails to attend three consecutive meetings of the Committee without any sufficient cause or reasons;

iv. any vacancy in the membership caused by any reason mentioned in clauses (ii) and (iii) shall be filled by the Government for the unexpired portion of two years term.

v. travelling and daily allowance shall be payable to the non-official members of the Committee at the highest rate admissible to the Government servants of Group 'A' under the rules and orders made by the Central Government and for the time being in force.

Provided that the payment of travelling allowance and daily allowance to a member who is a Member of the Parliament or a Member of a State Legislature shall be regulated in accordance with the Salary, Allowances and Pension of Members of Parliament Act, 1954 (30 of 1954) or the respective provisions of law pertaining to the member of the concerned State Legislature.

5. Conduct of business of the Committee.-

i. The Chairperson shall call the meeting of the Committee whenever considered necessary, but not less than once in a month.

ii. The meeting of the Committee shall be held at New Delhi.

iii. In a case where the Chairperson is satisfied that inspection of site or sites of forest land proposed to be used for non-forest purposes shall be necessary or expedient in connection with the consideration of the proposal or proposals received under sub-rule (3) of rule 6, he may direct that the meetings of the Committee to be held at a place other than New Delhi from where such inspection of site or sites is necessary.

iv. The Chairperson shall preside over every meeting of the Committee at which he is present.

v. Every question upon which the Central Government is required to be advised shall be considered in the meeting of the Committee provided that in urgent cases if the meeting cannot be convened within a month, the Chairperson may direct that papers may be circulated and sent to the members for their opinion within the stipulated time.

vi. The quorum of the meeting of the Committee shall be three.

6. Submission of the proposals seeking approval of the Central Government under section 2 of the Act.-

(1) Every user agency, who wants to use any forest land for non-forest purposes shall make his proposal in the appropriate Form appended to these rules, i.e. Form 'A' for proposals seeking first time approval under the Act and Form 'B' for proposals seeking renewal of leases where approval of the Central Government under the Act had already been obtained earlier, to the concerned nodal officer authorized in this behalf by the State Government, alongwith requisite information and documents, complete in all respects, well in advance of taking up any non-forest activity on the forest land.

(2) Every State Government or other authority, after having received the proposal under sub-rule (1) and after being satisfied that the proposal requires prior approval under section 2 of the Act, shall send the proposal to the Central Government in the appropriate forms, within ninety days of the receipt of the proposal from the user agency for proposals seeking first time approval under the Act and within sixty days for proposals seeking renewal of leases where approval of the Central Government under the Act had already been obtained earlier:

Provided that all proposals involving clearing naturally grown trees in forest land or portion thereof for the purpose of using it for reafforestation shall be sent in the form of Working Plan or Management Plan.

(3) The proposal referred to in sub-rule (2) above, involving forest land of more than forty hectare shall be sent by the State Government to the Secretary to the Government of India, Ministry of Environment and Forests, Paryavaran Bhavan, CGO Complex, Lodhi Road, New Delhi-110 003, with a copy of the proposal (with complete enclosures) to the concerned Regional Office.

(4) The proposal referred to in sub-rule (2) above, involving forest land up to forty hectare shall be sent to the Chief Conservator of Forests or Conservator of Forests of the concerned Regional Office of the Ministry of Environment and Forests.

(5) The proposal referred to in sub-rule (2) above, involving clearing of naturally grown trees in forest land or portion thereof for the purpose of using it for reafforestation shall be sent to the Chief Conservator of Forests or Conservator of Forests of the concerned Regional Office of the Ministry of Environment and Forests.

7. Committee to advise on proposals received by the Central Government.-

(1) The Central Government shall refer every proposal, complete in all respects, received by it under sub-rule (3) of rule 6 including site inspection report, wherever required, to the Committee for its advice thereon.

(2) The Committee shall have due regard to all or any of the following matters while tendering its advice on the proposals referred to it under sub-rule (1), namely:-

a. Whether the forests land proposed to be used for non-forest purpose forms part of a nature reserve, national park wildlife sanctuary, biosphere reserve or forms part of the habitat or any endangered or threatened species of flora and fauna or of an area lying in severely eroded catchment;

b. Whether the use of any forest land is for agricultural purposes or for the rehabilitation of persons displaced from their residences by reason of any river valley or hydro-electric project ;

c. Whether the State Government or the other authority has certified that it has considered all other alternatives and that no other alternatives in the circumstances are feasible and that the required area is the minimum needed for the purpose; and

d. Whether the State Government or the other authority undertakes to provide at its cost for the acquisition of land of an equivalent area and afforestation thereof.

(3) While tendering the advice, the Committee may also suggest any conditions or restrictions on the use of any forest land for any non-forest purpose, which in its opinion, would minimise adverse environmental impact.

8. Action of the Central Government on the advice of the Committee.–

The Central Government shall, after considering the advice of the Committee tendered under rule 7 and after such further enquiry as it may consider necessary, grant approval to the proposal with or without conditions or reject the same within sixty days of its receipt.

9. Proceedings against persons guilty of offences under the Act.-

(1) The Central Government may, by notification, authorize any officer not below the rank of Conservator of Forests or the concerned forest officer having territorial jurisdiction over the forest land in respect of which the said offence is said to have been committed, to file complaints against the person (s) prima-facie found guilty of offence under the Act or the violation of the rules made thereunder, in the court having jurisdiction in the matter.

Provided that no complaint shall be filed in the court, without giving the person (s) or officer (s) or authority (s) against whom the allegations of offence exist, an opportunity to explain his or their conduct and to show cause, by issuing a notice in writing of not less than sixty days, as to why a complaint should not be filed in the court against him or them for alleged offences.

(2) The officer authorised by the Central Government in sub-rule (1) may require any State Government or its officer or any person or any other authority to furnish to it within a specified period any reports, documents, statistics and any other information related to contravention of the Act or the rules made thereunder, considered necessary for making a complaint in any court of jurisdiction and every such State Government or officer or person or authority shall be bound to do so.

Form for seeking prior approval under section 2 of the proposals by the State Governments and other authorities

PART-I

(to be filled up by user agency)

1. Project details:

(i) Short narrative of the proposal and project/scheme for which the forest land is required.

(ii) Map showing the required forest land, boundary of adjoining forest on a 1:50,000 scale map.

(iii) Cost of the project:

(iv) Justification for locating the project in forest area.

(v) Cost-benefit analysis (to be enclosed).

(vi) Employment likely to be generated.

2. Purpose-wise break-up of the total land required:

3. Details of displacement of people due to the project, if any:

i. Number of families.

ii. Number of Scheduled Castes/Scheduled Tribe families

iii. Rehabilitation plan. (to be enclosed)

4. Whether clearance under Environment (Protection) Act, 1986 required? (Yes/No).

5. Undertaking to bear the cost of raising and maintenance of compensatory afforestation and/or penal compensatory

afforestation as well as cost for protection and regeneration of Safety Zone, etc. as per the scheme prepared by the State Government (undertaking to be enclosed).

6. Details of Certificates/documents enclosed as required under the instructions.

(To be filled by the concerned Deputy Conservator of Forests)

7. Location of the project/Scheme:

i. State/Union Territory

ii. District.

iii. Forest Division

iv. Area of forest land proposed for diversion (in ha.)

v. Legal status of forest

vi. Density of vegetation.

vii. Species-wise (scientific names) and diameter class-wise enumeration of trees (to be enclosed. In case of irrigation / hydel projects enumeration at FRL, FRL-2 meter & FRL-4 meter also to be enclosed.)

viii. Brief note on vulnerability of the forest area to erosion.

ix. Approximate distance of proposed site for diversion from boundary of forest.

x. Whether forms part of National Park, wildlife sanctuary, biosphere reserve, tiger reserve, elephant corridor, etc. (If so, the details of the area and comments of the Chief Wildlife Warden to be annexed).

xi. Whether any rare/endangered/unique species of flora and fauna found in the area- if so details thereof.

xii. Whether any protected archaeological/heritage site/ defence establishment or any other important monument is located in the area. If so, the details thereof with NOC from competent authority, if required.

8. Whether the requirement of forest land as proposed by the user agency in col. 2 of Part-I is unavoidable and barest minimum for the project. If no, recommended area item-wise with details of alternatives examined.

9. Whether any work in violation of the Act has been carried out (Yes/No). If yes, details of the same including period of work done, action taken on erring officials. Whether work in violation is still in progress.

10. Details of compensatory afforestation scheme:

i. Details of non forest area/degraded forest area identified for compensatory afforestation, its distance from adjoining forest, number of patches, size of each patch.

ii. Map showing non-forest/degraded forest area identified for compensatory afforestation and adjoining forest boundaries.

iii. Detailed compensatory afforestation scheme including species to be planted, implementing agency, time schedule, cost structure, etc.

iv. Total financial outlay for compensatory afforestation scheme.

v. Certificates from competent authority regarding suitability of area identified for compensatory afforestation and from management point of view. (To be signed by the concerned Deputy Conservator of Forests).

11. Site inspection report of the DCF (to be enclosed) especially highlighting facts asked in col. 7 (xi, xii), 8 and 9 above.

12. Division/District profile:

i. Geographical area of the district.

ii. Forest area of the district.

iii. Total forest area diverted since 1980 with number of cases.

iv. Total compensatory afforestation stipulated in the district/division since 1980 on (a) forest land including penal compensatory afforestation,

(b) non-forest land.

v. Progress of compensatory afforestation as on (date) ________________ on

(a) forest land

(b) non-forest land.

13. Specific recommendations of the DCF for acceptance or otherwise of the proposal with reasons.

Signature

Name

Official Seal

Date:-____________

Place:-____________

PART-III

(TO BE FILLED BY THE CONCERNED CONSERVATOR OF FORESTS)

14. Whether site, where the forest land involved is located has been inspected by concerned Conservator of Forests (Yes/No). If yes, the date of inspection & observations made in form of inspection note to be enclosed.

15. Whether the concerned Conservator of Forests agree with the information given in Part-B and the recommendations of Deputy Conservator of Forests.

16. Specific recommendation of concerned Conservator of Forests for acceptance or otherwise of the proposal with detailed reasons.

Signature

Name

Official Seal

Date:-____________

Place:-____________

PART-IV

(TO BE FILLED IN BY THE NODAL OFFICER OR PRINCIPAL CHIEF CONSERVATOR OF FORESTS OR HEAD OF FOREST DEPARTMENT)

17. Detailed opinion and specific recommendation of the State Forest Department for acceptance of otherwise of the proposal with remarks.

(While giving opinion, the adverse comments made by concerned Conservator of Forests or Deputy Conservator of Forests should be categorically reviewed and critically commented upon).

Signature

Name & Designation

(Official Seal)

Date:-______________

Place:-______________

PART- V

(TO BE FILLED IN BY THE SECRETARY IN CHARGE OF FOREST DEPARTMENT OR BY ANY OTHER AUTHORISED OFFICER OF THE STATE GOVERNMENT NOT BELOW THE RANK OF AN UNDER SECRETARY)

18. Recommendation of the State Government:

(Adverse comments made by any officer or authority in Part-B or Part-C or Part-D above should be specifically commented upon)

Signature

Name & Designation

(Official Seal)

Date:-______________

Place:-______________

INSTRUCTIONS (for Part-I):-

1. The project authorities may annex a copy of the approved project/plan in addition to filling Col. 1 (i) e.g. IBM approved mining plan for major minerals/CMPDI plan with subsidence analysis reports, etc.

2. Map has to be in original duly authenticated jointly by project authorities and concerned DCF – Col. 1 (ii).

3. Complete details of alternative alignments examined especially in case of project like roads, transmission lines, railway lines, canals, etc. to be shown on map with details of area of forest land involved in each alternative to be given - Col. 1 (iii).

4. For proposals relating to mining, certificate from competent authority like District Mining Officer about non-availability of the same mineral in surrounding/nearby non-forest areas.

5. In case the same company/individual has taken forest land for similar project in the State, a brief detail of all such approvals/leases be given as an enclosure along with current status of the projects.

6. The latest clarifications issued by the Ministry under Forest (Conservation) Act, 1980 may be kept in mind. In case such information do not fit in the given columns, the same shall be annexed separately.

GENERAL INSTRUCTIONS:-

1. On receipt of proposal, Nodal Officer shall issue a receipt to the user agency indicating therein the name of the proposal, user agency, area in hectare, serial number and date of receipt.

2. If the space provided above is not sufficient to specify any information, please attach separate details/documents.

3. While forwarding the proposal to the Central Government, complete details on all aspects of the case as per Form prescribed above read with the clarifications issued by the Ministry of Environment and Forests, Government of India, New Delhi should be given. Incomplete or deficient proposals shall not be

considered and shall be returned to the State Government in original.

4. The State Government shall submit the proposal to the Central Government within stipulated time limits. In case of delay while forwarding, the reasons for the same to be given in the forwarding/covering letter.

9

Forest Conservation Rules

(as amended up to May 1992)

G.S.R. 719... in exercise of the powers conferred by sub-section (i) of section 4 of the Forest (Conservation) Act, 1980 (69 of 1980), the Central Government hereby makes the following rules, namely

1. Short title, extent and commencement

(a) These rules may called the Forest (Conservation) Rules, 1981.

(b) They shall extend to the whole of India except the State of Jammu and Kashmir.

(c) They shall come into force on the date of their publication in the Official Gazette.

2. Definitions :-In these rules, unless the context otherwise requires -

(a) "Act" means the Forest (Conservation) Act, 1980 (69 of 1980)

(b) "Committee" means the Committee constituted under section 3

(c) "Chairman" means the Chairman of the Committee

(d) "Member" means a member of the Committee

(e) "Section" means a section of the Act.

2A (1) Composition of the Committee:- The Committee shall be composed of the following Members :

(i) Inspector General of Forests - Chairman
Ministry of Environment and Forests

(ii) Additional Inspector General of Forests - Member
Ministry of Environment and Forests

(iii) Joint Commissioner (Soil Conservation) - Member
Ministry of Agriculture

(iv) Three eminent environmentalists - Member
(non-officials)

(V) Deputy Inspector General of Forests - Member-Secretary
(Forest Conservation), Ministry of Environment and Forests

(2) Additional Inspector General of Forests shall act as the Chairman in the absence of Inspector General of Forests.

2B. Terms of appointment of non-official members shall be as follows

(i) A non-official member shall hold his office for a period of two years.

(ii) A non-official member shall cease to hold off ice if he dies, resigns, becomes of unsound mind, becomes insolvent or is convicted by a court of law on a criminal offence involving moral turpitude.

(iii) Any vacancy in the membership caused by any reason mentioned in sub-rule (ii) shall be filled by the Government for the unexpired portion of 2 years term.

(iv) Travelling and daily allowances shall be payable to the non-official members of the Committee at the highest rate admissible to the Government servants of Group W under the rules and orders made by the Central Government and for the time being in force.

Provided that the payment of travelling allowance and daily allowance to a member who is a Member of the Parliament or a Member of a State Legislature shall be regulated in accordance with the Salary, Allowance and Pension of Members of Parliament Act, 1954 or the respective provisions of law pertaining to the Members of the concerned State Legislature.

3. Conduct of business of the Committee

(1) The Chairman shall call the meeting of the Committee as often as necessary, but not less frequently than once in a month.

(2) The meetings of the Committee shall normally be held at New Delhi. However, in a case where the Chairman is satisfied that inspection of site or sites of forest land proposed to be used for non-forest purposes would be necessary or expedient in connection with the consideration of the proposal or proposals received under sub-rule (1) of rule-4, he may direct that the meetings of the Committee will be held at a place other than Delhi from where such inspection of she or sites can be undertaken conveniently.

(3) The Chairman shall preside over every meeting of the Committee at which he is present :

Provided that K the Chairman is absent from a meeting and it is not expedient to adjourn the meeting, the senior-most member of the Committee shall preside over the meeting.

(4) Every question upon which the Committee is required to advise shall be considered at its meeting provided that in urgent cases K the meeting of the Committee cannot be convened within a month, the Chairman may direct that necessary papers may be sent to the Members for their opinion by a stipulated date.

(5) The quorum for the meeting of the Committee shall be three.

4.

(1) Every State Government or other authority seeking the prior approval under section 2 shall send its proposal to the Central Government in the form appended to these rules :

Provided that all proposals involving clearing naturally grown trees in forest land or portion thereof for the purpose of using K for reafforestation shall be sent in the form of Working Plan / Management Plan.

(2) Every proposal referred to in sub-rule (1) shall be sent to the following address, namely :

Secretary to the Government of India

Ministry of Environment & Forests

Paryavaran Bhawan, CGO Complex

Lodi Road, New Delhi - 110003

Provided that all proposals involving forest land up to twenty hectares and proposals involving clearing of naturally grown trees in forest land or portion thereof for the purpose of using it for reafforestation shall be sent to the Chief Conservator of Forests 1 Conservator of Forests of the concerned Regional Office of the Ministry of Environment and Forests.

5. Committee to advise on proposals received by the Central Government (1) The Central Government shall refer every proposal received by it under sub-rule

(1) of rule 4 to the Committee for its advice thereon if the area of the forest land involved is more than twenty hectares. Provided that proposals involving clearing of naturally grown trees in forest land or portion thereof for the purpose of using it for reafforestation shall not be referred to the Committee for its advice.

(2) The Committee shall have due regard to all or any of the following matters while tendering its advice on the proposals referred to R under sub-rule (1), namely :

(a) Whether the forests land proposed to be used for non-forest purpose forms part of a nature reserve, national park wildlife sanctuary, biosphere reserve or forms part of the habitat of any endangered or threatened species of flora and fauna or of an area lying in severely eroded catchment;

(b) Whether the use of any forest land is for agricultural purposes or for the rehabilitation or persons displaced from their residences by reason of any river valley or hydro-electric project;

(c) Whether the State Government or the other authority has certified that it has considered all other alternatives and that no other alternatives in the circumstances are feasible and that the required area is the minimum needed for the purpose; and

(d) Whether the State Government or the other authority undertakes to provide at its cost for the acquisition of land of an equivalent area and afforestation thereof.

(3) While tendering the advice, the Committee may also suggest any conditions or restrictions on the use of any forest land for any non-forest purpose which, in its opinion, would minimise adverse environmental impact.

6. Action of the Central Government on the advice of the Committee -The Central Government shall. after considering the advice of the Committee tendered under rule 5 and after such further enquiry as it may consider necessary, grant approval to the proposal with or without conditions or reject the same.

FORM

Form for seeking prior approval under section 2 of the proposals by the State Governments and other authorities. (See rule 4)

1. Project details:

(i) Short narrative of the proposal and project 1 scheme for which the forest land is required.

(ii) Map showing the required forest area, boundary of adjoining forest and itemwise break-up of the required forest area for different purposes (to be authenticated by an officer not below the rank of Deputy Conservator of Forests).

(iii) Total cost of the project:

(iv) Justification for locating the project in the forest area giving alternatives examined and reasons for their rejection.

(v) Financial and social benefits.

(vi) Total population benefited.

(vii) Employment generated.

2. Location of the project 1 Scheme:

(i) State/Union Territory.

(ii) District.

(iii) Forest Division, Forest Block, compartment etc.

3. Item-wise break-up of the total land required for the project/scheme along with its existing land use.

4. Details of forest land involved:

(i) Legal status of the forest (namely, reserve, protected/unclassed, etc.)

(ii) Details of flora and fauna existing in the area.

(iii) Density of vegetation.

(iv) Species-wise and diameter class-wise abstract of trees.

(v) Vulnerability of the forest area to erosion, whether it forms a part of a seriously eroded area or not.

(vi) Whether h forms a part of national park, wildlife sanctuary, nature reserve, biosphere reserve, etc; and 9 so, details of the area involved. (Specific comments of the Chief Wildlife Warden to be annexed).

(vii) Item-wise break-up of the forest land required for the project/scheme for different purposes.

(viii) Rare/endangered species of flora and fauna found in the area.

(ix) Whether it is a habitat for migrating fauna or forms a breeding ground for them.

(X) Any other significance of the area relevant to the proposal.

5. Details of displacement of people due to the project :

(i) Total number of families involved in displacement.

(ii) Number of Scheduled Caste/Scheduled Tribe families involved in displacement.

(iii) Detailed rehabilitation plan.

6. Details of compensatory afforestation scheme:

(i) Details of non forest area/degraded forest area identified for compensatory afforestation, its distance from adjoining forest, number of patches, size of each patch.

(ii) Map showing non-forest/degraded forest area identified for compensatory afforestation an adjoining forest boundaries.

(iii) Detailed compensatory afforestation scheme including species to be planted, implementing agency, time schedule, cost structure etc.

(iv) Total financial outlay, for compensatory afforestation scheme.

(v) Certificates from competent authority regarding suitability of area identified for compensatory afforestation for afforestation and from management point of view. (To be signed by an officer not below the rank of Deputy Conservator of Forests).

(vi) Certificate from the Chief Secretary regarding non-availability of the nonforest land for compensatory afforestation (if applicable).

7. Details regarding Transmission Lines (only for Transmission Line proposals):

(i) Total length of the Transmission Line.

(ii) Length passing through forest area.

(iii) Right of Way.

(iv) Number of Towers to be erected.

(v) Number of Towers to be erected in forest area.

(vi) Height of Transmission Towers.

8. Details of Irrigation Hydel Project (only for Irrigation/ Hydel Projects):

(i) Total catchment area.

(ii) Total command area.

(iii) Full Reservoir Level.

(iv) High Flood Level.

(v) Minimum Drawal Level.

(vi) Break-up of area failing in catchment area of the project (forest land, cultivated land, pasture land, human cultivation, others).

(vii) Area of submergence at High Flood Level.

(viii) Area of submergence at Full Reservoir Level.

(ix) Area of submergence at 2 metre below Full Reservoir Level.

(x) Area of submergence at 4 metre below Full Reservoir Level (For medium and major projects only).

(xi) Area of submergence at Minimum Drawal Level.

(xii) Detailed catchment area treatment plan.

(xiii) Total financial outlays and details regarding availability of funds for Catchment Area Treatment Plan.

9. Details regarding Road/Railway Lines (only for Roads/ Railway lines proposal):

(i) Length and width of the strip and forest area required.

(ii) Total length of the road.

(iii) Length of the road already constructed

(iv) Length of the road passing through the forest.

10. Details regarding mining proposals (only for mining proposals):

(i) Total mining lease area and forest area required.

(ii) Period of mining lease proposed.

(iii) Estimated reserve of each mineral/ore in the forest area and in the non-forest area.

(iv) Annual estimated production of mineral/ore.

(v) Nature of mining operations (open cast/underground)

(vi) Phased reclamation plan.

(vii) Gradient of the area where mining would be undertaken.

(viii) Copy of the Lease Deed (to be attached only for renewal purposes).

(ix) Number of labourers to be employed.

(x) Area of forest land required for

(a) Mining.

(b) Storing mineral/ore.

(c) Dumping of overburden.

(d) Storing tools and machinery.

(e) Construction of building power stations, workshops, etc.

(f) Township/housing colony

(g) Construction of road/rope way/railway lines.

(h) Full land use plan of forest area required.

(xi) Reasons why any of the activities, referred to in (a) to (h) above under the project for which forest land has been asked for cannot be undertaken/located outside forest area.

(xii) The extent of damage likely to be caused and the number of trees affected on account of mining and related activities.

(xiii) Distance of the mining area from perennial watercourses, national and State highways, national parks, sanctuaries and biosphere reserves.

(xiv) Procedure for stocking of the topsoil for reuse.

(xv) Extent of subsidence expected in underground mining operations and its impact on water, forest and other vegetation.

11. Cost-benefit analysis.

12. Whether clearance from environmental angle is required (Yes/NO).

If, yes, whether, requisite details for the same have been furnished (Yes/No

13. Whether any work in violation of the Act has been carded out (Yes/No).

If yes,

(i) Details of the same including date of commencement.

(ii) Officers responsible for violation of the Act.

(iii) Action taken/being taken against erring officers.

(iv) Whether work in violation of the Act is still in progress.

14. Any other information....

15. Details of Certificates/documents enclosed.

16. Detailed opinion of the Chief Conservator of Forests/ Head of the Forest Department concerned covering the following aspects, namely :

(i) Out-turn of timer, fuelwood and other forest produce from the forest land involved.

(ii) Whether the district is self-sufficient in timber and fuelwood, and

(iii) The effect of the proposal on

(a) Fuelwood supply to rural population

(b) Economy and livelihood of the tribals and backward communities.

(iv) Specific recommendations of the Chief Conservator of Forests/head of the forest Department for acceptance or otherwise of the proposal with reasons thereof.

Certified that all other alternatives for the purpose have been explored and the demand for the required area is the minimum demand for forest land.

MINISTRY OF ENVIRONMENT & FORESTS BIOLOGICAL DIVERSITY ACT 2002 AND ESTABLISHMENT OF NATIONAL BIODIVERSITY AUTHORITY ,CHENNAI

India is party to the Convention on Biological Diversity(CBD) 1992 which recognizes the sovereign rights of states to use their own Biological Resources. In order to help in realizing the objectives of CBD, India has enacted an umbrella legislation called the biological Diversity Act 2002(No.18 of 2003) aimed at conservation of biological resources and associated knowledge as well as facilitating access to them in a sustainable manner and through a just process.

In exercise of the powers conferred by Sub-Section (1) (4) of Section 8 of the Biological Diversity Act, 2002 (18 of 2003), the Central Government has established a body called the National Biodiversity Authority, on and from the 1st day of October, 2003.

The main functions of the Authority are :

a) To lay down procedures and guidelines to govern the activities provided under Section 3, 4, and 6. (Permission to foreigners/NRI's foreign companies)

- **For obtaining any biological resource (Section -3)**
- **For transferring the results of any research (Section -4).**

Certain collaborative research projects exempted (Section 5)

b) To advice the Government of India. Specific areas mentioned as per the Act are the following:

i) Notifications of threatened species (Section -38)

ii) Designate institutions as repositories for different categories of biological resources (Section -39)

iii) Exempt certain biological resources, normally traded as commodities (Section -40)

c) To encourage setting up State Biodiversity Boards

d) To build up database and documentation system

e) To creat awareness through mass media

i. **Training of personnel**

ii. **Necessary measures in the areas of Intellectual Propriety** Rights

MINISTRY OF LAW AND JUSTICE
(LEGISLATIVE DEPARTMENT)

New Delhi, the 5th February, 2003/Magha 16, 1924 (Saka)

The following Act of Parliament received the assent of the President on the 5th February, 2003, and is hereby published for general information:

THE BIOLOGICAL DIVERSITY ACT, 2002

No. 18 OF 2003

[5th February, 2003]

An Act to provide for conservation of biological diversity, sustainable use of its components and fair and equitable sharing of the benefits arising out of the use of biological resources, knowledge and for matters connected therewith or incidental thereto.

WHEREAS India is rich in biological diversity and associated traditional and contemporary knowledge system relating thereto.

AND WHEREAS India is a party to the United Nations Convention on Biological Diversity signed at Rio de Janeiro on the 5th day of June, 1992;

AND WHEREAS the said Convention came into force on the 29th December, 1993;

AND WHEREAS the said Convention reaffirms the sovereign rights of the States over their biological resources;

AND WHEREAS the said Convention has the main objective of conservation of biological diversity, sustainable use of its components and fair and equitable sharing of the benefits arising out of utilization of genetic resources;

AND WHEREAS it is considered necessary to provide for conservation, sustainable utilization and equitable sharing of the benefits arising out of utilization of genetic resources and also to give effect to the said Convention.

BE it enacted by Parliament in the Fifty third Year of the Republic of India as follows:–

PRELIMINARY

1. Short title, extent and com-mencement

(1) This Act may be called the Biological Diversity Act, 2002.

(2) It extends to the whole of India.

(3) It shall come into force on such date as the Central Government may, by notification in the Official Gazette, appoint:

Provided that different dates may be appointed for different provisions of this Act and any reference in any such provision to the commencement of this Act shall be construed as a reference to the coming into force of that provision.

2. Definitions

In this Act, unless the context otherwise requires,–

(a) "benefit claimers" means the conservers of biological resources, their byproducts, creators and holders of knowledge and information relating to the use of such biological resources, innovations and practices associated with such use and application;

(b) "biological diversity" means the variability among living organisms from all sources and the ecological complexes of which they are part, and includes diversity within species or between species and of eco systems;

(c) "biological resources" means plants, animals and micro organisms or parts thereof, their genetic material and by products (excluding value added products) with actual or potential use or value, but does not include human genetic material;

(d) "bio survey and bio utilization" means survey or collection of species, subspecies, genes, components and extracts of biological resource for any purpose and includes characterization, inventorisation and bioassay;

(e) "Chairperson" means the Chairperson of the National Biodiversity Authority or, as the case may be, of the State Biodiversity Board;

(f) "commercial utilization" means end uses of biological resources for commercial utilization such as drugs, industrial enzymes, food flavours, fragrance, cosmetics, emulsifiers, oleoresins, colours, extracts and genes used for improving crops and livestock through genetic intervention, but does not include conventional breeding or traditional practices in use in any agriculture, horticulture, poultry, dairy farming, animal husbandry or bee keeping;

(g) "fair and equitable benefit sharing" means sharing of benefits as determined by the National Biodiversity Authority under section 2 1;

(h) "local bodies" means Panchayats and Municipalities, by whatever name called, within the meaning of clause (1) of article 243B and clause (1) of article 243Q of the Constitution and in the absence of any Panchayats or Municipalities, institutions of self government constituted under any other provision of the Constitution or any Central Act or State Act;

(i) "member" means a member of the National Biodiversity Authority or a State Biodiversity Board and includes the Chairperson;

(j) "National Biodiversity Authority" means the National Biodiversity Authority established under section 8;

(k) "prescribed" means prescribed by rules made 1 under this Act;

(l) "regulations" means regulations made under this Act;

(m) "research" means study or systematic investigation of any biological resource or technological application, that uses biological systems, living organisms or derivatives thereof to make or modify products or processes for any use;

(n) "State Biodiversity Board" means the State Biodiversity Board established under section 22;

(o) "sustainable use" means the use of components of biological diversity in such manner and at such rate that does not lead to the long term decline of the biological diversity thereby maintaining its potential to meet the needs and aspirations of present and future generations;

(p) "value added products" means products which may contain portions or extracts of plants and animals in unrecognizable and physically inseparable form.

REGULATION OF ACCESS TO BIOLOGICAL DIVERSITY

Certain persons not to undertake Biodiversity related activities without approval of National Biodiversity Authority

3. (1) No person referred to in sub section (2) shall, without previous approval of the National Biodiversity Authority, obtain any biological resource occurring in India or knowledge associated thereto for research or for commercial utilization or for bio survey and bio utilization.

(2) The persons who shall be required to take the approval of the National Biodiversity Authority under sub section (1) are the following, namely:

(a) a person who is not a citizen of India;

(b) a citizen of India, who is a non resident as defined in clause (30) of section 2 of the Income tax Act, 196 1;

(c) a body corporate, association or organization-

(i) not incorporated or registered in India; or

(ii) incorporated or registered in India under any law for the time being in force which has any non Indian participation in its share capital or management.

Results of research not to be transferred to certain persons without approval of National Biodiversity Authority

4. No person shall, without the previous approval of the National Biodiversity Authority, transfer the results of any research relating to any biological resources occurring in, or obtained from, India for monetary consideration or otherwise

to any person who is not a citizen of India or citizen of India who is non resident as defined in clause (30) of section 2 of the Income tax Act, 1961 or a body corporate or organization which Is not registered or incorpo-rated in India or which has any non Indian participation in its share capital or management.

Explanation. For the purposes of this section, "transfer" does not include publication of research papers or dissemination of knowledge in any seminar or workshop, if such publication is as per the guidelines issued by the Central Government.

Sections 3 and 4 not to apply to certain collaborative research projects

5.(1) The provisions of sections 3 and 4 shall not apply to collaborative research projects involving transfer or exchange of biological resources or information relating thereto between institutions, including Government sponsored institutions of India, and such institutions in other countries, if such collaborative research projects satisfy the conditions specified in sub section (3).

(2) All collaborative research projects, other than those referred to in sub section (1) which are based on agreements concluded before the commencement of this Act and in force shall, to the extent the provisions of agreement are inconsistent with the provisions of this Act or any guidelines issued under clause (a) of sub section (3), be void.

(3) For the purposes of sub section (1), collaborative research projects shall-

(a) conform to the policy guidelines issued by the Central Government in this behalf;

(b) be approved by the Central Government.

Application for intellectual property rights not to be made without approval of National Biodiversity Authority

6.(1) No person shall apply for any intellectual property right, by whatever name called, in or outside India for any invention based on any research or information on a biological resource obtained from India without obtaining the previous approval of the National Biodiversity Authority before making such application.

Provided that if a person applies for a patent, permission of the National Biodiversity Authority may be obtained after the acceptance of the patent but before the seating of tile patent by the patent authority concerned:

Provided further that the National Biodiversity Authority shall dispose of the application for permission made to it within a period of ninety days from the date of receipt thereof.

(2) The National Biodiversity Authority may, while granting the approval under this section, impose benefit sharing fee or royalty or both or impose conditions including the sharing of financial benefits arising out of the commercial utilization of such rights.

(3) The provisions of this section shall not apply to any person making an application for any right under any law relating to protection of plant varieties enacted by Parliament.

(4) Where any right is granted under law referred to in sub section (3), the concerned authority granting such right shall endorse a copy of such document granting the right to the National Biodiversity Authority.

Prior intimation to State Biodiversity Board for obtaining biological resource for certain purposes

7. No person, who is a citizen of India or a body corporate, association or organization which is registered in India, shall obtain any biological resource for commercial utilization, or bio survey and bio utilization for commercial utilization except after giving prior intimation to the State Biodiversity Board concerned:

Provided that the provisions of this section shall not apply to the local people and communities of the area, including growers and cultivators of biodiversity, and *vaids* and *hakims*, who have been practicing indigenous medicine.

NATIONAL BIODIVERSITY AUTHORITY

Establishment of National Biodiversity Authority

8.(1) With effect from such date as the Central Government may, by notification in the Official Gazette, appoint, there shall be established by the Central Government for the pur-poses of this Act, a body to be called the National Biodiversity Authority.

(2) The National Biodiversity Authority shall be a body corporate by the name aforesaid, having perpetual succession and a common seal, with power to acquire, hold and dispose of property, both movable and immovable, and to contract, and shall by the said name sue and be sued.

(3) The head office of the National Biodiversity Authority shall be at Chennai and the National Biodiversity Authority may, with the previous approval of the Central Government, establish offices at other places in India.

(4) The National Biodiversity Authority shall consist of the following members,

(a) a Chairperson, who shall be an eminent person having adequate knowledge and experience in the conservation and sustainable use of biological diversity and in matters relating to equitable sharing of benefits, to be appointed by the Central Government;

(b) three ex officio members to be appointed by the Central Government, one representing the Ministry dealing with Tribal Affairs and two representing the Ministry dealing with Environment and Forests of whom one shall be the Additional Director General of Forests or the Director General of Forests;

(c) seven ex officio members to be appointed by the Central Government to represent respectively the Ministries of the Central Government dealing with –

(i) Agricultural Research and Education;

(ii) Biotechnology;

(iii) Ocean Development;

(iv) Agriculture and Cooperation;

(v) Indian Systems of Medicine and Homoeopathy;

(vi) Science and Technology;

(vii) Scientific and Industrial Research;

(d) five non official members to be appointed from amongst specialists and scientists having special knowledge of, or experience in, matters relating to conservation of biological diversity, sustainable use of biological resources and equitable sharing of benefits arising out of the use of biological resources, representatives of industry, conservers, creators and know ledge holders of biological resources.

9. The term of office and conditions of service of the Chairperson and the other members other than *ex officio* members of the National Biodiversity Authority shall be such as may be prescribed by the Central Government.

Chairperson to be Chief Executive of National Biodiversity Authority

10. The Chairperson shall be the Chief Executive of the National Biodiversity Authority and shall exercise such powers and perform such duties, as may be prescribed.

Removal of Members

11. The Central Government may remove from the National Biodiversity Authority any member who, in its opinion, has-

(a) been adjudged as an insolvent; or

(b) been convicted of an offence which involves moral turpitude; or

(c) become physically or mentally incapable of acting as a member; or

(d) so abused his position as to render his continuance in office detrimental to the public interest; or

(e) acquired such financial or other interest as is likely to affect prejudicially his functions as a member.

Meetings of National Biodiversity Authority

12. (1) The National Biodiversity Authority shall meet at such time and place and shall observe such rules of procedure in regard to the transaction of business at its meetings (including the quorum at its meetings) as may be prescribed.

(2) The Chairperson of the National Biodiversity Authority shall preside at the meetings of the National Biodiversity Authority.

(3) If for any reason the Chairperson is unable to attend any meeting of the National Biodiversity Authority, any member of the National Biodiversity Authority chosen by the members present at the meeting shall preside at the meeting.

(4) All questions which come before any meeting of the National Biodiversity Authority shall be decided by a majority of votes of the members present and voting and in the event of equality of votes, the Chairperson or, in his absence, the person presiding, shall have and exercise a second or casting vote.

(5) Every member who is in any way, whether directly, indirectly or personally, concerned or interested in a matter to be decided at the meeting shall disclose the nature of his concern or interest and after such disclosure, the member concerned or interested shall not attend that meeting.

(6) No act or proceeding of the National Biodiversity Authority shall be invalidated merely by reason of-

(a) any vacancy in, or any defect in the constitution of, the National Biodiversity Authority; or

(b) any defect in the appointment of a person acting as a member; or

(c) any irregularity in the procedure of the National Biodiversity Authority not affecting the merits of the case.

Committees of National Biodiversity Authority

13. (1) The National Biodiversity Authority may constitute a committee to deal with agro biodiversity.

Explanation For the purposes of this sub section, "agro biodiversity" means bio-logical diversity, of agriculture related species and their wild relatives.

(2) Without prejudice to the provisions of sub section (1), the National Biodiversity Authority may constitute such number of committees as it deems fit for the efficient dis-charge of its duties and performance of its functions under this Act.

(3) A committee constituted under this section shall co-opt such number of persons, who are not the members of the National Biodiversity Authority, as it may think fit and the persons so co opted shall have the right to attend the meetings of the committee and take part in its proceedings but shall not have the right to vote.

(4) The persons appointed as members of the committee under sub section (2) shall be entitled to receive such allowances or fees for attending the meetings of the committee as may be fixed by the Central Government.

Officers and employees of National Biodiversity Authority

14.(1) The National Biodiversity Authority may appoint such officers and other employees as it considers necessary for the efficient discharge of its functions under this Act.

(2) The terms and conditions of service of such officers and other employees of the National Biodiversity Authority shall be such as may be specified by regulations.

Authentication of Orders and Decisions of National Biodiversity Authority

15. All orders and decisions of the National Biodiversity Authority shall be authenticated by the signature of the Chairperson or any other member authorized by the National Biodiversity Authority in this behalf and all other instruments executed by the National Biodiversity Authority shall be authenticated by the signature of an officer of the National Biodiversity Authority authorized by it in this behalf.

Delegation of Powers

16. The National Biodiversity Authority may, by general or special order in writing, delegate to any member, officer of the National Biodiversity Authority or any other person subject to such conditions, if any, as may be specified in the order, such of the powers and functions under this Act (except the power to prefer an appeal under section 50 and the power to make regulations under section 64) as it may deem necessary.

Expenses of National Biodiversity Authority to be defrayed out of the Consolidated Fund of India

17. The salaries and allowances payable to the members and the administrative expenses of the National Biodiversity Authority including salaries, allowances and pension payable to, or in respect of, the officers and other employees of the National Biodiversity Authority shall be defrayed out of the Consolidated Fund of India.

FUNCTIONS AND POWERS OF THE NATIONAL BIODIVERSITY AUTHORITY

18.(1) It shall be the duty of the National Biodiversity Authority to regulate activities referred to in sections 3, 4 and 6 and by regulations issue guidelines for access to biological resources and for fair and equitable benefit sharing.

(2) The National Biodiversity Authority may grant approval for undertaking any activity referred to in sections 3, 4 and 6.

(3) The National Biodiversity Authority may-

(a) advise the Central Government on matters relating to the conservation of biodiversity, sustainable use of its components and equitable sharing of benefits arising out of the utilization of biological resources;

(b) advise the State Governments in the selection of areas of biodiversity importance to be notified under sub section (1) of section 37 as heritage sites and measures for the management of such heritage sites;

(c) perform such other functions as may be necessary to carry out the provisions of this Act.

(4) The National Biodiversity Authority may, on behalf of the Central Government, take any measures necessary to oppose the grant of intellectual property rights in any country outside India on any biological resource obtained from India or knowledge associated with such biological resource which is derived from India.

APPROVAL BY THE NATIONAL BIODIVERSITY AUTHORITY

19.(1) Any person referred. to in sub section (2) of section 3 who intends to obtain any biological resource occurring in India or knowledge associated thereto for research or for commercial utilization or for bio survey and bio utilization or transfer the results of any research relating to biological resources occurring in, or obtained from, India, shall make application in such form and payment of such fees as may be prescribed, to the National Biodiversity Authority.

(2) Any person who intends to apply for a patent or any other form of intellectual property protection whether in India or outside India referred to in sub section (1) of section 6, may make an application in such form and in such manner as may be prescribed to the National Biodiversity Authority.

(3) On receipt of an application under sub section (1) or sub section (2), the National Biodiversity Authority may, after making such enquiries as it may deem fit and if necessary after consulting an expert committee constituted for this purpose, by order, grant approval subject to any regulations made in this behalf and subject to such terms and conditions as it may deem fit, including the imposition of charges by way of royalty or for reasons to be recorded in writing, reject the application:

Provided that no such order for rejection shall be made without giving an opportunity of being heard to the person affected.

(4) The National Biodiversity Authority shall give public notice of every approval granted by it under this section.

Transfer of biological resource or knowledge.

20.(1) No person who has been granted approval under section 19 shall transfer any biological resource or knowledge associated thereto which is the subject matter of the said approval except with the permission of the National Biodiversity Authority.

(2) Any person who intends to transfer any biological resource or knowledge associated thereto referred to in sub section (1) shall make an application in such form and in such manner as may be prescribed to the National Biodiversity Authority.

(3) On receipt of an application under sub section (2), the National Biodiversity Authority may, after making such enquiries as it may deem fit and if necessary after consulting an expert committee constituted for this purpose, by order, grant approval subject to such terms and conditions as it may deem fit, including the imposition of charges by way of royalty or for reasons to be recorded in writing, reject the application:

Provided that no such order for rejection shall be made without giving an opportunity of being heard to the person affected.

(4) The National Biodiversity Authority shall give public notice of every approval granted by it under this section.

Determination of equitable benefit sharing by National Biodiversity Authority

21.(1) The National Biodiversity Authority shall while granting approvals under section 19 or section 20 ensure that the terms and conditions subject to which approval is granted secures equitable sharing of benefits arising out of the use of accessed biological resources, their by products, innovations and practices associated with their use and applications and knowledge relating thereto in accordance with mutually agreed terms and conditions between the person applying for such approval, local bodies concerned and the benefit claimers.

(2) The National Biodiversity Authority shall, subject to any regulations made in this behalf, determine the benefit

sharing which shall be given effect in all or any of the following manner, namely:-

(a) grant of joint ownership of intellectual property rights to the National Biodiversity Authority, or where benefit claimers are identified, to such benefit claimers;

(b) transfer of technology;

(c) location of production, research and development units in such areas which will facilitate better living standards to the benefit claimers;

(d) association of Indian scientists, benefit claimers and the local people with research and development in biological resources and bio survey and bio utilization;

(e) setting up of venture capital fund for aiding the cause of benefit claimers;

(f) payment of monetary compensation and non monetary benefits to the benefit claimers as the National Biodiversity Authority may deem fit.

(3) Where any amount of money is ordered by way of benefit sharing, the National Biodiversity Authority may direct the amount to be deposited in the National Biodiversity Fund:

Provided that where biological resource or knowledge was a result of access from specific individual or group of individuals or organizations, the National Biodiversity Authority may direct that the amount shall be paid directly to such individual or group of individuals or organizations in accordance with the terms of any agreement and in such manner as it deems fit.

(4) For the purposes of this section, the National Biodiversity Authority shall, in consultation with the Central Government, by regulations, frame guidelines.

STATE BIODIVERSITY BOARD

Establishment of State Biodiversity Board

22.(1) With effect from such date as the State Government may, by notification in the Official Gazette, appoint in this behalf, there shall be established by that Government for the purposes of this Act, a Board for the State to be known as the (name of the State) Biodiversity Board.

(2) Notwithstanding anything contained in this section, no State Biodiversity Board shall be constituted for a Union territory and in relation to a Union territory, the National Biodiversity Authority shall exercise the powers and perform the functions of a State Biodiversity Board for that Union territory:

Provided that in relation to any Union territory, the National Biodiversity Authority may delegate all or any of its powers or functions under this sub section to such person or group of persons as the Central Government may specify.

(3) The Board shall be a body corporate by the name aforesaid, having perpetual succession and a common seal, with power to acquire, hold and dispose of property, both movable and immovable, and to contract, and shall by the said name sue and he sued.

(4) The Board shall consist of the following members, namely:

(a) a Chairperson who shall be an eminent person having adequate knowledge and experience in the conservation and sustainable use of biological diversity and in matters relating to equitable sharing of benefits, to he appointed by the State Government;

(b) not more than five ex officio members to be appointed by the State Government to represent the concerned Departments of the State Government;

(c) not more than five members to be appointed from amongst experts in matters relating to conservation of biological diversity, sustainable use of biological resources and equitable sharing of benefits arising out of the use of biological resources.

(5) The head office of the State Biodiversity Board shall be at such place as the State Government may, by notification in the Official Gazette, specify.

Functions of State Biodiversity Board

23. The functions of the State Biodiversity Board shall be to–

(a) advise the State Government, subject to any guidelines issued by the Central Government, on matters relating to the conservation of biodiversity, sustainable use of its components and equitable sharing of the benefits arising out of the utilization of biological resources;

(b) regulate by granting of approvals or otherwise requests for commercial utilization or bio survey and bio utilization of any biological resource by Indians;

(c) perform such other functions as may he necessary to carry out the provisions of this Act or as may be prescribed by the State Government.

Power of State Biodiversity Board to restrict certain activities

24.(1) Any citizen of India or a body corporate, organization or association registered in India intending to undertake any activity referred to in section 7 shall give prior intimation in such form as may be prescribed by the State Government to the State Biodiversity Board.

(2) On receipt of an intimation under sub section (1), the State Biodiversity Board may, in consultation with the local bodies concerned and after making such enquires as it conservation, may deem fit, by order, prohibit or restrict any such activity if it is of opinion that such activity is detrimental or contrary to the objectives of conservation and sustainable use of biodiversity or equitable sharing of benefits arising out of such activity:

Provided that no such order shall be made without giving an opportunity of being heard to the person affected.

(3) Any information given in the form referred to in sub section (1) for prior intimation shall be kept confidential and shall not be disclosed, either intentionally or unintentionally, to any person not concerned thereto.

Provisions of sections 9 to 17 to apply with modifications to State Biodiversity Board

25. The provisions of sections 9 to 17 shall apply to a State Biodiversity Board and shall have effect subject to the following modifications, namely:

(a) references to the Central Government shall be construed as references to the State Government;

(b) references to the National Biodiversity Authority shall be construed as references to the State Biodiversity Board;

Provisions of sections 9 to 17 to apply with modifications to State Biodiversity Board.

(c) reference to the Consolidated Fund of India shall be construed as reference to the Consolidated Fund of the State.

FINANCE, ACCOUNTS AND AUDIT OF NATIONAL BIODIVERSITY AUTHORITY

Grants or loans by the Central Government

26. The Central Government may, after due appropriation made by Parliament by law in this behalf, pay to the National Biodiversity Authority by way of grants or loans such sums of money as the Central Government may think fit for being utilized for the purposes of this Act.

Constitution of National Biodiversity Fund

27.(1) There shall be constituted a Fund to be called the National Biodiversity Fund and there shall be credited thereto

(a) any grants and loans made to the National Biodiversity Authority under section 26;

(b) all charges and royalties received by the National Biodiversity Authority under this Act; and

(c) all sums received by the National Biodiversity Authority from such other sources as may be decided upon by the Central Government.

(2) The Fund shall be applied for—

(a) channeling benefits to the benefit claimers;

(b) conservation and promotion of biological resources and development of areas from where such biological resources or knowledge associated thereto has been accessed;

(c) socio economic development of areas referred to in clause (b) in consultation with the local bodies concerned.

Annual report of National Biodiversity Authority

28. The National Biodiversity Authority shall prepare, in such form and at such time each financial year as may be prescribed, its annual report, giving a full account of its activities during the previous financial year and furnish, to the Central Government, before such date as may be prescribed, its audited copy of accounts together with auditors report thereon.

Budget accounts and audit

29.(1) The National Biodiversity Authority shall prepare a budget, maintain proper accounts and other relevant records (including the accounts and other relevant records of the National Biodiversity Fund) and prepare an annual statement of account in such form as may be prescribed by the Central Government in consultation with the Comptroller and Auditor General of India.

(2) The accounts of the National Biodiversity Authority shall be audited by the Comptroller and Auditor General of India at such intervals as may be specified by him and any expenditure incurred in connection with such audit shall be payable by the National Biodiversity Authority to the Comptroller and Auditor General of India.

(3) The Comptroller and Auditor General of India and any other person appointed by him in connection with the audit of the accounts of the National Biodiversity Authority shall have the same rights and privileges and authority in connection with such audit as the Comptroller and Auditor General generally has in connection with the audit of the Government accounts and, in particular, shall have the right to demand the production of books, accounts, connected vouchers and other documents

and papers and to inspect any of the offices of the National Biodiversity Authority.

(4) The accounts of the National Biodiversity Authority as certified by the Comptroller and Auditor General of India or any other person appointed by him in this behalf together with the audit report thereon shall be forwarded annually to the Central Government.

Annual Report to be laid before parliament

30. The Central Government shall cause the annual report and auditor's report to be laid, as soon as may be after they are received, before each House of Parliament.

FINANCE, ACCOUNTS AND AUDIT OF STATE BIODIVERSITY BOARD

31. The State Government may, after due appropriation made by the State Legislature by law in this behalf, pay to the State Biodiversity Board by way of grants or loans such sums of money as the State Government may think fit for being utilized for the purposes of this Act.

Constitution of State Biodiversity Fund

32.(1) There shall be constituted a Fund to be called the State Biodiversity Fund and there shall be credited thereto

(a) any grants and loans made to the State Biodiversity Board under section 31;

(b) any grants or loans made by the National Biodiversity Authority;

(c) all sums received by the State Biodiversity Board from such other sources as may be decided upon by the State Government.

(2) The State Biodiversity Fund shall be applied for-

(a) the management and conservation of heritage sites;

(b) compensating or rehabilitating any section of the people economically affected by notification under sub section (1) of section 37;

(c) conservation and promotion of biological resources;

(d) socio economic development of areas from where such biological resources or knowledge associated thereto has been accessed subject to any order made under section 24, in consultation with the local bodies concerned;

(e) meeting the expenses incurred for the purposes authorized by this Act.

Annual report of State Biodiversity Board

33. The State Biodiversity Board shall prepare, in such form and at such time in each financial year as may be prescribed, its annual report, giving a full account of its activities during the previous financial year, and submit a copy thereof to the State Government.

Audit of accounts of State Biodiversity Board

34. The accounts of the State Biodiversity Board shall be maintained and audited in such manner as may, in consultation with the Accountant General of the State, be prescribed and the State Biodiversity Board shall furnish, to the State Government, before such date as may be prescribed, its audited copy of accounts together with auditor's report thereon.

Annual Report of State Biodiversity Board to be laid before State Legislature

35. The State Government shall cause the annual report and auditor's report to be laid, as soon as may be after they are received, before the House of State Legislature.

DUTIES OF THE CENTRAL AND THE STATE GOVERNMENTS

Central Government to Develop National Strategies Plans. Etc., for Conservation, etc., of Biological Diversity

36.(1) The Central Government shall develop national strategies, plans, programmes for the conservation and promotion and sustainable use of biological diversity including measures for identification and monitoring of areas rich in biological resources, promotion of *in situ*, and *ex situ*, conservation of biological resources, incentives for research, training and public education to increase awareness with respect to biodiversity.

(2) Where the Central Government has reason to believe that any area rich in biological diversity, biological resources and their habitats is being threatened by overuse, abuse or neglect, it shall issue directives to the concerned State Government to take immediate ameliorative measures, offering such State Government any technical and other assistance that is possible to be provided or needed.

(3) The Central Government shall, as far as practicable wherever it deems appropriate, integrate the conservation, promotion and sustainable use of biological diversity into relevant sectoral or cross sectoral plans, programmes and policies.

4) The Central Government shall undertake measures,-

(i) wherever necessary, for assessment of environmental impact of that project which is likely to have adverse effect on biological diversity, with a view to avoid or minimize such effects and where appropriate provide for public participation in such assessment;

(ii) to regulate, manage or control the risks associated with the use and release of living modified organisms resulting from biotechnology likely to have adverse impact on the conservation and sustainable use of biological diversity and human health.

(5) The Central Government shall endeavour to respect and protect the knowledge of local people relating to biological diversity, as recommended by the National Biodiversity Authority through such measures, which may include registration of such knowledge at the local, State or national levels, and other measures for protection, including *sui generis* system.

Explanation:-For the purposes of this section,–

(a) "*ex situ* conservation" means the conservation of components of biological diversity outside their natural habitats;

(b) "*in situ* conservation" means the conservation of ecosystems and natural habitats and the maintenance and recovery of viable populations of species in their natural surroundings and, in the case of domesticated or cultivated species, in the surroundings where they have developed their distinctive properties.

Biodiversity Heritage Sites

37.(1) Without prejudice to any other law for the time being in force, the State Government may, from time to time in consultation with the local bodies, notify in the Official Gazette, areas of biodiversity importance as biodiversity heritage sites under this Act.

(2) The State Government, in consultation with the Central Government, may frame rules for the management and conservation of all the heritage sites.

(3) The State Government shall frame schemes for compensating or rehabilitating any person or section of people economically affected by such notification.

Power of Central Government to Notify Threatened Species

38. Without prejudice to the provisions of any other law for the time being in force, the Central Government, in consultation with the concerned State Government, may from time o time notify any species which is on the verge of extinction or likely to become extinct in the near future as a threatened species and prohibit or regulate collection thereof for any purpose and take appropriate steps to rehabilitate and preserve those species.

Power of Central Government to Designate Repositories

39.(1) The Central Government may, in consultation with the National Biodiversity Authority, designate institutions as repositories under this Act for different categories of biological resources.

(2) The repositories shall keep in safe custody the biological material including voucher specimens deposited with them.

(3) Any new taxon discovered by any person shall be notified to the repositories or any institution designated for this purpose and he shall deposit the voucher specimens with such repository or institution.

Power of Central Government to exempt Certain Biological Resources

40. Notwithstanding anything contained in this Act, the Central Government may, in consultation with the National Biodiversity Authority, by notification in the Official Gazette, declare that the provisions of this Act shall not apply to any items, including biological resources normally traded as commodities.

BIODIVERSITY MANAGEMENT COMMITTEES

Constitution of Biodiversity Management Committee

41.(1) Every local body shall constitute a Biodiversity Management Committee within its area for the purpose of promoting conservation, sustainable use and documentation of biological diversity including preservation of habitats, conservation of land races, folk varieties and cultivars, domesticated stocks and breeds of animals and microorganisms and chronicling of knowledge relating to biological diversity.

Explanation. For the purposes of this sub section,—

(a) "cultivar" means a variety of plant that has originated and persisted under cultivation or was specifically bred for the purpose of cultivation;

(b) "folk variety" means a cultivated variety of plant that was developed, grown and exchanged informally among farmers;

(c) "landrace" means primitive cultivar that was grown by ancient farmers and their successors.

(2) The National Biodiversity Authority and the State Biodiversity Boards shall consult the Biodiversity Management Committees while taking any decision relating to the use of biological resources and knowledge associated with such resources occurring within the territorial jurisdiction of the Biodiversity Management Committee.

(3) The Biodiversity Management Committees may levy charges by way of collection fees from any person for accessing or collecting any biological resource for commercial purposes from areas falling within its territorial jurisdiction.

LOCAL, BIODIVERSITY FUND

Grants to Local Biodiversity Fund

42. The State Government may, after due appropriation made by State Legislature by law in this behalf, pay to the Local Biodiversity Funds by way of grants or loans such sums of money as the State Government may think fit for being utilized for the purposes of this Act.

Constitution of Local Biodiversity Fund

43.(1) There shall be constituted a Fund to be called the Local Biodiversity Fund at every area notified by the State Government where any institution of self government is functioning and there shall be credited thereto-

(a) any grants and loans made under section 42;

(b) any grants or loans made by the National Biodiversity Authority;

(c) any grants or loans made by the State Biodiversity Boards;

(d) fees referred to in sub section (3) of section 41 received by the Biodiversity Management Committees;

(e) all sums received by the Local Biodiversity Fund from such other sources as may be decided upon by the State Government.

Application of Local Biodiversity Fund

44.(1) Subject to the provisions of sub section (2), the management and the custody of the Local Biodiversity Fund and the purposes for which such Fund shall be applied, be in the manner as may be prescribed by the State Government.

(2) The Fund shall be used for conservation and promotion of biodiversity in the areas falling within the jurisdiction of the

concerned local body and for the benefit of the community in so far such use is consistent with conservation of biodiversity.

Annual Report of Biodiversity Management Committees

45. The person holding the custody of the Local Biodiversity Fund shall prepare, in such form and during each financial year at such time as may be prescribed, its annual report, giving a full account of its activities during the previous financial year, and submit a copy thereof to the concerned local body.

Audit of accounts of Biodiversity Management Committees

46. The accounts of the Local Biodiversity Fund shall be maintained and audited in such manner as may, in consultation with the Accountant General of the State, be prescribed and the person holding the custody of the Local Biodiversity Fund shall furnish, to the concerned local body, before such date as may be prescribed, its audited copy of accounts together with auditor's report thereon.

Annual report, etc, of the Biodiversity Management Committee to be submitted to district Magistrate

47. Every local body constituting a Biodiversity Management Committee under sub section (1) of section 41, shall cause, the annual report and audited copy of accounts together with auditor's report thereon referred to in sections 45 and 46, respectively and relating to such Committee to be submitted to the District Magistrate having jurisdiction over the area of the local body.

MISCELLANEOUS

National Biodiversity Authority to be Bound by the Directions given by Central Government

48.(1) Without prejudice to the foregoing provisions of this Act, the National Biodiversity Authority shall, in the discharge of its functions and duties under this Act, be bound by such directions on questions of policy as the Central Government may give in writing to it from time to time.

Provided that the National Biodiversity Authority shall, as far as practicable, be given opportunity to express its views before any direction is given under this sub section.

(2) The decision of the Central Government whether a question is one of policy or not shall be final.

Power of State Government to give Directions

49.(1) Without prejudice to the foregoing provisions of this Act, the State Biodiversity Board shall, in the discharge of its functions and duties under this Act, be bound by such directions on questions of policy as the State Government may give in writing to it from time to time:

Provided that the State Biodiversity Board shall, as far as practicable, be given an opportunity to express its views before any direction is given under this sub section.

(2) The decision of the State Government whether a question is one of policy or not shall be final.

Settlement of disputes between State Biodiversity Boards

50.(1) If a dispute arises between the National Biodiversity Authority and a State Biodiversity Board, the said Authority or the Board, as the case may be, may prefer an appeal to the Central Government within such time as ma be prescribed.

(2) Every appeal made under sub section (1) shall be in such form as may be pre-scribed by the Central Government.

(3) The procedure for disposing of an appeal shall be such as may be prescribed by the Central Government:

Provided that before disposing of an appeal, the parties shall be given a reasonable opportunity of, being heard.

(4) If a dispute arises between the State Biodiversity Boards, the Central Government shall refer the same to the National Biodiversity Authority.

(5) While adjudicating any dispute under sub section (4), the National Biodiversity Authority shall be guided by the principles of natural justice and shall follow such proce-dure as may be prescribed by the Central Government.

(6) The National Biodiversity Authority shall have, for the purposes of discharging its functions under this section, the same powers as are vested in a civil court under the Code of Civil Procedure, 1908 in respect of the following matters, namely–

(a) summoning and enforcing the attendance of any person and examining him on oath;

(b) requiring the discovery and production of documents;

(c) receiving evidence on affidavits;

(d) issuing commissions for the examination of witnesses or documents;

(e) reviewing its decisions; dismissing an application for default or deciding it *ex parte*;

(g) setting aside any order of dismissal of any application for default or any order passed by it *ex parte*;

(h) any other matter which may be prescribed.

(7) Every proceeding before the National Biodiversity Authority shall be deemed to be a judicial proceeding within the meaning of sections 193 and 228, and for the purpose of section 196, of the Indian Penal Code and the National Biodiversity Authority shall be deemed to be a civil court for all the purposes of section 195 and Chapter XXV1 of the Code of Criminal Procedure, 1973.

Members, officers, etc., of National Biodiversity Authority and State Biodiversity Board deemed to be public servants

51. All members, officers and other employees of the National Biodiversity Authority or the State Biodiversity Board shall be deemed, when acting or purporting to act in pursuance of any of the provisions, of this Act, to be public servants within the meaning of section 21 of the Indian Penal Code.

Appeal

52. Any person, aggrieved by any determination of benefit sharing or order of the National Biodiversity Authority or a State Biodiversity Board under this Act, may file an appeal to

the High Court within thirty days from the date of communication to him, of the determination or order of the National Biodiversity Authority or the State Biodiversity Board, as the case may be:

Provided that the High Court may, if it is satisfied that the appellant was prevented by sufficient cause from filing the appeal within the said period, allow it to be filed within a further period not exceeding sixty days.

Execution of Determination or Order

53. Every determination of benefit sharing or order made by the National Biodiversity Authority or a State Biodiversity Board under this Act or the order made by the High Court in any appeal against any determination or order of the National Biodiversity Authority or a State Biodiversity Board shall, on a certificate issued by any officer of the National Biodiversity Authority or a State Biodiversity Board or the Registrar of the High Court, as the case may be, be deemed to be decree of the civil court and shall be executable in the same manner as a decree of that court.

Explanation. For the purposes of this section and section 52, the expression "State Biodiversity Board" includes the person or group of persons to whom the powers or functions under sub section (2) of section 22 have been delegated under the proviso to that sub--section and the certificate relating to such person or group of persons under this section shall be issued by such person or group of persons, as the case may be.

Protection of Action taken in Good Faith

54. No suit, prosecution or other legal proceedings shall lie against the Central Government or the State Government or any officer of the Central Government or the State Government or any member, officer or employee of the National Biodiversity Authority or the State Biodiversity Board for anything which is in good faith done or intended to he done under this Act or the rules or regulations made thereunder.

Penalties

55.(1) Whoever contravenes or to or abets the contravention of the provisions of section 3 or section 4 or section 6 shall be punishable with imprisonment for a term which may extend to five years, or with fine which may extend to ten lakh rupees and where the damage caused exceeds tend lakh rupees such fine may commensurate with the damage caused, or with both.

(2) Whoever contravenes or attempts to contravene or abets the contravention of the provisions of section 7 or any order made under sub section (2) of section 24 shall be punishable with imprisonment for a term which may extend to three years, or with fine which may extend to five lakh rupees, or with both.

Penalty for contravention of directions or orders of Central government, State government, National Biodiversity Authority and State Biodiversity Boards

56. If any person contravenes any direction given or order made by the Central Government, the State Government, the National Biodiversity Authority or the State Biodiversity Board for which no punishment has been separately provided under this Act, he shall be punished with a fine which may extend to one lakh rupees and in case of a second or subsequent offence, with fine which may extend to two lakh rupees and in the case of continuous contravention with additional fine which may extend to two lakh rupees everyday during which the default continues.

Offences by Companies

57.(1) Where an offence or contravention under this Act has been committed by a company, every person who at the time the offence or contravention was committed was in charge of, and was responsible to, the company for the conduct of the business of the company, as well as the company, shall be deemed to be guilty of the offence or contraven-tion and shall be liable to be proceeded against and punished accordingly:

Provided that nothing contained in this sub section shall render any such person liable to any punishment provided in this Act, if he proves that the offence or contravention was committed without his knowledge or that he had exercised all due diligence to prevent the commission of such offence or contravention.

(2) Notwithstanding anything contained in sub section (1), where an offence or con-travention under this Act has been committed by a company and it is proved that the offence or contravention has been committed with the consent or connivance of, or is attributable to, any neglect on the part of any director, manager, secretary or other officer of the company, such director, manager, secretary or other officer shall also be deemed to be guilty of the offence or contravention and shall be liable to be proceeded against and punished accordingly.

Explanation. For the purposes of this section,-- (a) "company" means any body corporate and includes a firm or other associa-tion of individuals; and (b) "director", in relation to a firm, means a partner in the firm.

Offences to be Cognizable and non Bailable

58. The offences under this Act shall be cognizable and non bailable.

Act to have Effect in Addition to other Acts

59. The provisions of this Act shall be in addition to, and not in derogation of, the provisions in any other law, for the time being in force, relating to forests or wildlife.

Power of Central Government to giver Directions to State Government

60. The Central Government may give directions to any State Government as to the carrying into execution in the State of any of the provisions of this Act or of any rule or regulation or order made thereunder.

Cognizance of Offences

61. No Court shall take cognizance of any offence under this Act except on a com-plaint made by--

(a) the Central Government or any authority or officer authorized in this behalf by that Government; or

(b) any benefit claimer who has given notice of not less than thirty days in the prescribed manner, of such offence and of his intention to make a complaint, to the Central Government or the authority or officer authorized as aforesaid.

Power of Central Government to make Rules

62.(1) The Central Government may' by notification in the Official Gazette, make rules for carrying out the purposes of this Act.

(2) In particular, and without prejudice to the generality of the foregoing power, such rules may provide for A or any of the following matters, namely:–

(a) terms and conditions of service of the Chairperson and members under section 9;

(b) powers and duties of the Chairperson under section 10;

(c) procedure under sub section (1) of section 12 in regard to transaction of business at meetings,

(d) form of application and payment of fees for undertaking certain activities under sub-section (1) of section 19;

(e) the form and manner of making an application under sub section (2) of section 19;

(f) form of application and the manner for transfer of biological resource or knowledge under sub section (2) of section 20;

(g) form in which, and the time of each financial year at which, the annual report of the National Biodiversity Authority shall be prepared and the date before which its audited copy of accounts together with auditor's report thereon shall be furnished under section 28;

(h) form in which the annual statement of account shall be prepared under subsection (1) of section 29;

(i) the time within which and the form in which, an appeal may be preferred, the procedure for disposing of an appeal and the procedure for adjudication, under section 50;

(j) the additional matter in which the National Biodiversity Authority may exercise powers of the civil court under clause (h) of sub section (6) of section 50;

(k) the manner of giving notice under clause (b) of section 61;

(1) any other matter which is to be, or may be, prescribed, or in respect of which provision is to be made, by rules.

(3) Every rule made under this section and every regulation made under this Act shall be laid, as soon as may he after it is made, before each House of Parliament, while it Is in session or a total period of thirty days which may be comprised in one session or in two or more successive sessions, and if, before the expiry of the session immediately following. The session or the successive sessions aforesaid, both Houses agree in making any modification in the rule or regulation or both Houses agree that the rule or regulation should not be made, the rule or regulation shall thereafter have effect only in such modified form or be of no effect, as the case may be; so, however, that any such modification or annulment shall be without prejudice to the validity of anything previously done under that rule or regulation.

Power of State Government to make Rules

63.(1) The State Government may, by notification in the Official Gazette, make rules for carrying out the purposes of this Act. (2) In particular, and without prejudice to the generality of the foregoing power, such rules may provide for all or any of the following matters, namely:– (a) the other functions to be performed by the State Biodiversity Board under clause (c) of section 23; (b) the form in which the prior intimation shall be given under sub section (1) of section 24; (c) the form in which, and the time of each financial year at which, the annual report shall be prepared under section 33; (d) the manner of maintaining and auditing the accounts of the State Biodiversity Board and the date before which its audited copy of the accounts together with auditor's report thereon shall be furnished under section 34; (e) management and conservation of national heritage sites under section 37; (f) the manner of management and custody of the Local Biodiversity Fund and the purposes for which such Fund shall be applied under sub section (1) of section 44; (g) the form of annual report and the time at which such report shall be prepared during each financial year under section 45; (h) the manner of maintaining and auditing the accounts of the Local Biodiversity Fund and the date before which its audited copy of the accounts together with auditor's report thereon shall be furnished under section 46; (i) any other matter which is to be, or may be, specified. (3) Every rule made by the State Government under this section shall he laid, as soon as may be after it is made, before each House of the State Legislature where it consists of two Houses, or where such Legislature consists of one House, before that House.

Power of Remove Difficulties

64. The National Biodiversity Authority shall, with the previous approval of the Central Government, by notification in the Official Gazette, make regulations for carrying out the purposes of this Act.

Power to Remove Difficulties

65. (1) If any difficulty arises in giving effect to the provisions of this Act, the Central Government may, by order, not inconsistent with the provisions of this Act, remove the difficulty:

Provided that no such order shall be made after the expiry of a period of two years from the commencement of this Act.

(2) Every order made under this section shall be laid, as soon as may be, after it is made, before each House of Parliament.

10

THE WILD LIFE PROTECTION AMENDMENT

NO. 16 OF 2003

[*17thJanuary, 2003.*]

An Act further to amend the Wild Life (Protection) Act, 1972.

BE it enacted by Parliament in the Fifty third Year of the Republic of India as follows:

Short Title and Commencement

1. (1) This Act may be called the Wild Life (Protection) Amendment Act, *2002.*

(2) It shall come into force on such date as the Central Government may, by notification in the Official Gazette, appoint and different dates may be appointed for different provisions of this Act.

Amendment of Long Title

53 of 1972

2. In the Wild Life (Protection) Act, 1972 (hereinafter referred to as the principal Act), for the long title, the following long title shall be substituted, namely:–

"An Act to provide for the protection of wild animals, birds and plants and for matters connected therewith or ancillary or incidental thereto with a view to ensuring the ecological and environmental security of the country.".

Amendment of Section 2

3. In section 2 of the principal Act,– *(a)* for clause *(1)*, the following clause shall be substituted, namely: - '*(1)* "animal" includes mammals, birds, reptiles, amphibians, fish, other chordates and invertebrates and also includes their young and eggs;';

(b) for clause *(4)*, the following clause shall be substituted, namely: - '*(4)* "Board" means a State Board for Wild Life constituted under sub-section *(1)* of section 6;';

(c) clause *(8)* shall be omitted;

(d) for clause *(9)*, the following clause shall be substituted, namely: - '*(9)* "Collector" means the chief officer in charge of the revenue administration of a district or any other officer not below the rank of a Deputy Collector as may be appointed by the State Government under section 18B in this behalf;';

(e) for clause *(11)*, the following clause shall be substituted, namely:– '*(11)* "dealer" in relation to any captive animal, animal article, trophy, uncured trophy, meat or specified plant, means a person, who carries on the business of buying or selling any such animal or article, and includes a person who undertakes business in any single transaction;';

(f) for clause *(12A)*, the following clauses shall be substituted, namely:—

16 of 1927

'*(12A)* "Forest officer" means the Forest officer appointed under clause *(2)* of section 2 of the Indian Forest Act, 1927 or under any other Act for the time being in force in a State;

16 of 1927

(12B) "forest produce" shall have the same meaning as in sub clause *(b)* of clause *(4)* of section 2 of the Indian Forest Act, 1927;';

(g) in clause *(16)*, for sub clauses *(a)* and *(b)*, the following sub clauses shall be substituted, namely:-

"*(a)* killing or poisoning of any wild animal or captive animal and every attempt to do so;

(b) capturing, coursing, snaring, trapping, driving or baiting any wild or captive animal and every attempt to do so;";

(h) for clause *(18A)*, the following clause shall be substituted namely:–

'*(18A)* "livestock" means farm animals and includes buffaloes, bulls, bullocks, camels, cows, donkeys, goats, sheep, homes, mules, yaks, pigs, ducks, geese, poultry and their young but does not include any animal specified in Schedules I to V;';

(i) for clauses *(19)* and *(20)*, the following clauses shall be substituted, namely: -

'*(19)* "manufacturer" means a person who manufactures articles from any animal or plant specified in Schedules I to V and VI, as the case may be;

(20) "meat" includes blood, bones, sinew, eggs, shell or carapace, fat and flesh with or without skin, whether raw or cooked, of any wild animal or captive animal, other than a vermin;

(20A) "National Board" means the National Board for Wild Life constituted under section 5A;';

(j) after clause *(24)*, the following clause shall be inserted, namely:—

'*(24A)* "protected area means a National Park, a sanctuary, a conservation reserve or a community reserve notified under sections 18, 35, 36A and 36C of the Act;';

(k) for clauses *(25B)* and *(26)*, the following clauses shall be substituted, namely:—

'*(25B)* "reserve forest" means the forest declared to be reserved by the State Government under section 20 of the Indian Forest Act, 1927, or declared as such under any other State Act;

(26) "sanctuary" means an area declared as a sanctuary by notification under the provisions of Chapter IV of this Act and

shall also include a deemed sanctuary under sub section (4) of section 66;';

(l) clause *(28)* shall be omitted;

(m) for clause *(30)*, the following clause shall be substituted, namely:—

'*(30)* "taxidermy", with its grammatical variations and cognate expressions, means the curing, preparation or preservation or mounting of trophies;';

(n) in clause *(31)*, for sub clause *(b)*, the following sub clause shall be substituted, namely:—

"*(b)* antler, bone, carapace, shell , horn, rhinoceros horn, hair, feather, nail, tooth, tusk, musk, eggs, nests and honeycomb;";

(o) for clause *(36)*, the following clause shall be substituted, namely:—

'*(36)* "wild animal" means any animal specified in Schedules I to IV and found wild in nature;';

(p) for clause *(37)*, the following clause shall be substituted, namely:—

'*(37)* "wild life" includes any animal, aquatic or land vegetation which forms part of any habitat;';

(q) in clause *(39)*, for the words "but does not include a circus arid an establishment", the words "and includes a circus and rescue centres but does not include an establishment" shall be substituted.

Amendment of Section 3

4. In section 3 of the principal Act, –

(i) in sub section *(1)*, clause *(b)* shall be omitted;

(ii) for sub section *(3)*, the following sub section shall be substituted, namely:—

"*(3)* The officers and other employees appointed under this section shall be required to assist the Director.".

Amendment of Section 4

5. In section 4 of the principal Act, in sub section *(1)*, for clause *(bb)*, the following clause shall be substituted, namely:–

"*(bb)* Honorary Wild Life Wardens;".

Insertion of New Sections 5A to 5C

6. After section 5 of the principal Act, the following sections shall be inserted, namely:

Constitution of the National Board for Wild Life

"5A. *(1)* The Central Government shall, within three months from the date of commencement of the Wild Life (Protection) Amendment Act, 2002, constitute the National Board for Wild Life consisting of the following members, namely:

(a) the Prime Minister as Chairperson;

(b) the Minister in charge of Forests and Wild Life as Vice Chairperson;

(c) three members of Parliament of whom two shall be from the House of the People and one from the Council of States;

(d) Member, Planning Commission in charge of Forests and Wild Life;

(e) five persons to represent non governmental organisations to be nominated by the Central Government;

(f) ten persons to be nominated by the Central Government from amongst eminent conservationists, ecologists and environmentalists;

(g) the Secretary to the Government of India in charge of the Ministry or Department of the Central Government dealing with Forests and Wild Life;

(h) the Chief of the Army Staff,

(i) the Secretary to the Government of India in charge of the Ministry of Defence;

(j) the Secretary to the Government of India in charge of the Ministry of information and Broadcasting;

(k) the Secretary to the Government of India in; charge of the Department of Expenditure, Ministry of Finance;

(l) the Secretary to the Government of India, Ministry of Tribal Welfare;

(m) the Director General of Forests in the Ministry or Department of the Central Government dealing with Forests and Wild Life;

(n) the Director General of Tourism, Government of India;

(o) the Director General, Indian Council for Forestry Research and Education, Dehradun;

(p) the Director, Wild Life Institute of India, Dehradun;

(q) the Director, Zoological Survey of India;

(r) the Director, Botanical Survey of India;

(s) the Director, Indian Veterinary Research Institute;

(t) the Member Secretary, Central Zoo Authority;

(u) the Director, National Institute of Oceanography;

(v) one representative each from ten States and Union territories by rotation, to be nominated by the Central Government;

(w) the Director of Wild Life preservation who shall be the Member Secretary of the National Board.

(2) The term of office of the members other than those who are members ex officio, the manner of filling vacancies referred to in clauses *(e)*, *(1)* and *(v)* of subsection *(1)*, and the procedure to be followed in the discharge of their functions by the members of the National. Board shall be such, as may be prescribed.

(3) The members (except members *ex officio*) shall be entitled to receive such allowances in respect of expenses incurred in the performance of their duties as may be prescribed.

(4) Notwithstanding anything contained in any other law for the time being in force, the office of the member of the National Board shall not be deemed to be an office of profit.

Standing Committee of the National Board

5B. *(1)* The National Board may, in its discretion, constitute a Standing Committee for the purpose of exercising such powers and performing such duties as may be delegated to the Committee by the National Board.

(2) The Standing Committee shall consist of the Vice Chairperson, the Member--Secretary, and not more than ten members to be nominated by the Vice Chairperson from amongst the members of the National Board.

(3) The National Board may constitute committees, sub committees or study groups, as may be necessary, from time to time in proper discharge of the functions assigned to it.

Functions of the National Board

5C. *(1)* It shall be the duty of the National Board to promote the conservation and development of wild life and forests by such measures as it thinks fit.:

(2) Without prejudice to the generality of the foregoing provision, the measures referred to therein may provide for--

(a) framing policies and advising the Central Government and the State Governments on the ways and means of promoting wild life conservation and effectively controlling poaching and illegal trade of wild life and its products;

(b) making recommendations on the setting up of and management of national parks, sanctuaries and other protected areas and on matters relating to restriction of activities in those areas;

(c) carrying out or causing to be carried but impact assessment of various projects and activities on wild life or its habitat;

(d) reviewing from time to time, the progress in the field of wild life conservation in the country and suggesting measures for improvement thereto; and

(e) preparing and publishing a status report at least once in two years on wild life in the country.".

Substitution of New Section for Section 6

7. For section 6 of the principal Act, the following section shall be substituted, namely:

Constitution of State Board for Wild Life

"6. *(1)* The State Government shall, within a period of six months from the date of commencement of the Wild Life (Protection) Amendment Act, 2002 constitute a State Board for Wild Life consisting of the following members, namely:—

(a) the Chief Minister of the State and in case of the Union territory, either Chief Minister or Administrator, as the case may be – Chairperson;

(b) the Minister in charge of Forests and Wild Life – Vice Chairperson;

(c) three members of the State Legislature or in the case of a Union territory with Legislature, two members of the Legislative Assembly of that Union territory;

(d) three persons to represent non governmental organisations dealing with wild life to be nominated by the State Government;

(e) ten persons to be nominated by the. State Government from amongst eminent conservationists, ecologists and environmentalists including at least two representatives of the Scheduled Tribes;

(f) the Secretary to the State Government or the Government of the Union territory, as the case may be, in charge of Forests and Wild Life;

(g) the Officer in charge of the State Forest Department;

(h) the Secretary to the State Government, Department of Tribal Welfare;

(i) the Managing Director, State Tourism Development Corporation;

(j) an officer of the State Police Department not below the rank of Inspector- General;

(k) a representative of the Armed Forces not below the rank of a Brigadier to be nominated by the Central Government;

(l) the Director, Department of Animal Husbandry of the State;

(m) the Director, Department of Fisheries of the State;

(n) an officer to be nominated by the Director, Wild We Preservation;

(o) a representative of the Wild Life Institute of India, Dehradun;

(p) a representative of the Botanical Survey of India;

(q) a representative of the Zoological Survey of India;

(r) the Chief Wild Life Warden, who shall be the Member Secretary.

(2) The term of office of the members other than those who are members *ex officio* and the manner of filling vacancies referred to in clauses *(d)* and *(e)* of sub- section *(1)* and procedure to be followed shall be such, as may be prescribed.

(3) The member (except members *ex officio* shall be entitled to receive such allowances in respect of expenses incurred in the performance of their duties as may be prescribed.".

Amendment of Section 8

8. In section 8 of the principal Act,—

(i) for the words 'qhe Wild Life Advisory Board", the words "State Board for Wild Life" shall be substituted;

(ii) for clause *(a)*, the following clause shall be substituted, namely:—

"*(a)* in the selection and management of areas to be declared as protected areas;".

Amendment of Section 11

9. In section 11 of the principal Act, in sub section *(1)*,—

(i) in clause *(a)*, the following provisos and *Explanation* thereunder shall be inserted, namely:—

"Provided that no wild animal shall be ordered to be killed unless the Chief Wild Life Warden is satisfied that such animal cannot be captured, tranquilised or translocated:

Provided further that no such captured animal shall be kept in captivity unless the Chief Wild Life Warden is satisfied that such animal cannot be rehabilitated in the wild and the reasons for the same are recorded in writing.

Explanation–For the purposes of clause (a), the process or capture or translocation, as the case may be, of such animal shall be made in such manner as to cause minimum trauma to the said animal.";

(ii) in clause *(b),* for the words "such animal or cause such animal to be hunted", the words "such animal or group of animals in a specified area or cause such animal or group of animals in that specified area to be hunted" shall be substituted.

Substitution of Heading of Chapter IV

10. In Chapter IV of the principal Act, for the heading "SANCTUARIES, NATIONAL PARK AND CLOSED AREAS", the following heading shall be substituted, namely:

"PROTECTED AREAS"

Insertion of new sections 18A and 18B

11. After section 18 of the principal Act, the following sections shall be inserted, namely:—

Protection to Sanctuaries

"18A. *(1)* When the State Government declares its intention under sub section of section 18 to constitute any area, not comprised within any reserve forest or territorial waters under that sub section, as a sanctuary, the provisions of sections 27 to 33A (both inclusive) shall come into effect forthwith.

(2) Till such time as the rights of affected persons are finally settled under sections 19 to 24 (both inclusive), the State Government shall make alternative arrangements required for making available fuel, fodder and other forest produce to the

persons affected in terms of their rights as per the Government records.

Appointment of Collector

18B. The State Government shall appoint, an officer to act as Collector under the Act, within ninety days of coming into force of the Wild Life (Protection) Amendment Act, 2002, or within thirty days of the issue of notification under section 18, to inquire into and determine the existence, nature and extent of rights of any person in or over the land comprised within the limits of the sanctuary which may be notified under sub-section *(1)* of section 18.".

Amendment of Section 21

12. In section 21 of the principal Act, after the words "the Collector shall", the words "within a period of sixty days," shall be inserted.

Insertion of New Section 25A

13. After section 25 of the principal Act, the following section shall be inserted, namely:

Time-limit for Completion Acquisition Proceedings

"25A. *(1)*The Collector shall, as far as possible, complete the proceedings under sections 19 to 25 (both inclusive), within a period of two years from the date of notification of declaration of sanctuary under section 18.

(2) The notification shall not lapse if, for any reasons, the proceedings are not completed within a period of two years.".

Amendment of Section 26

14. In section26A of the principal Act for sub section *(3)*, the following sub section shall be substituted, namely:

"*(3)* No alteration of the boundaries of a sanctuary shall be made by the State Government except on a recommendation of the National Board.".

Substitution of New Section for Section 29

15. For section 29 of the principal Act, the following section shall be substituted, namely:–

Destruction etc., in a Sanctuary Prohibited without a Permit

"29. No person shall destroy, exploit or remove any wild life including forest produce from a sanctuary or destroy or damage or divert the habitat of any wild animal by any act whatsoever or divert, stop or enhance the flow of water into or outside the sanctuary, except under and in accordance with a permit granted by the Chief Wild Life Warden, and no such permit shall be granted unless the State Government being satisfied in consultation with the Board that such removal of wild life from the sanctuary or the change in the flow of water into or outside the sanctuary is necessary for the improvement and better management of wild life therein, authorises the issue of such permit:

Provided that where the forest produce is removed from a sanctuary the same may be used for meeting the personal bonafide needs of the people living in and around the sanctuary and shall not be used for any commercial purpose.

Explanation For the purposes of this section, grazing or movement of livestock permitted under clause *(d)* of section 33 shall not be deemed to be an act prohibited under this section.".

Amendment of Section 33

16. In section 33 of the principal Act after clause *(a)*, the following proviso shall be inserted, namely:–

"Provided that no construction of commercial tourist lodges, hotels, zoos and safari parks shall be undertaken inside a sanctuary except with the prior approval of the National Board;".

Insertion of New Section 33B

17. After section 33A of the principal Act, the following section shall be inserted, namely:–

"33B. *(1)* The State Government shall constitute an Advisory Committee consisting of the Chief Wild Life Warden or his nominee not below the rank of Conservator of Forests as its head and shall include a member of the State Legislature within whose constituency the sanctuary is situated, three representatives of Panchayati Raj Institutions, two representatives of non governmental organisations and three individuals active in the field of wild life conservation, one representative each from departments dealing with Home and Veterinary matters. Honorary Wild Life Warden, if any, and the officer in charge of the sanctuary as Member Secretary.

(2) The Committee shall render advice on measures to be taken for better conservation and management of the sanctuary including participation of the people living within and around the sanctuary.

(3) The Committee shall regulate its own procedure including quorum.".

Insertion of New Section 34A

18. After section 34 of the principal Act, the following section shall be inserted,

namely :–

Power to remove Encroachment

"34A., *(1)* Notwithstanding anything contained in any other law for the time being in force, any officer not below the rank of an Assistant Conservator of Forests may,–

(a) evict any person from a sanctuary or National Park, who unauthorisedly occupies Government land in contravention of the provisions of this Act;

(b) remove any unauthorised structures, buildings, or constructions erected on any Government land within any sanctuary or National Park and all the things, tools and effects belonging to such person shall be confiscated, by an order of an officer not below the rank of the Deputy Conservator of Forests:

Provided that no such order shall be passed unless the affected person is given an opportunity of being heard.

(2) The provisions of this section shall apply notwithstanding any other penalty which may be inflicted for violation of any other provision of this Act.".

Amendment of Section 35

19. In section 35 of the principal Act,–

(i) for sub sections *(5)* and *(6)*, the following sub sections shall be substituted, namely:–

"*(5)* No alteration of the boundaries of a National Park by the State Government shall be made except on a recommendation of the National Board.

(6) No person shall destroy, exploit or remove any Wild Life including forest produce from a National Park or destroy or damage or divert the habitat of any wild animal by any act whatsoever or divert, stop or enhance the flow of water into or outside the National Park, except under and in accordance with a permit granted by the Chief Wild Life Warden; and no such permit shall be granted unless the State Government being satisfied in consultation with the National Board that such removal of wild life from the National Park or the change in the flow of water into or outside the National Park is necessary for the improvement and better management of wild life therein, authorises the issue of such permit:

Provided that where the forest produce is removed from a National Park, the same may be used for meeting the personal *bona fide* needs of the people living in and around the National Park and shall not be used for any commercial purpose.";

(ii) after sub section *(8)*, the following *Explanation* shall be inserted, namely:–

"*Explanation.* For the purposes of this section, in case of an area, whether within a sanctuary or not, where the rights have been extinguished and the land has become vested in the State Government under any Act or otherwise, such area may be notified by it, by a notification, as a National Park and the proceedings under sections 19 to 26 (both inclusive) and the provisions of sub sections *(3)* and *(4)* of this section shall not apply.".

Insertion of New Sections 36A to 36D

20. After section 36 of the principal Act, the following sections shall be inserted, namely:

Declaration and Management of a Conservation Reserve

"36A. *(1)* The State Government may, after having consultations with the local communities, declare any area owned by the Government, particularly the areas adjacent to National Parks and sanctuaries and those areas which link one protected area with another, as a conservation reserve for protecting landscapes, seascapes, flora and fauna and their habitat:

Provided that where the conservation reserve includes any land owned by the Central Government, its prior concurrence shall be obtained before making such declaration.

(2) The provisions of sub section *(2)* of section 18, sub sections *(2)*, *(3)* and *(4)* of section 27, sections 30, 32 and clauses *(b)* and *(c)* of section 33 shall, as far as may be, apply in relation to a conservation reserve as they apply in relation to a sanctuary

Conservation Reserve Management Committee

36B. *(1)* The State Government shall constitute a conservation reserve management committee to advise the Chief Wild Life Warden to conserve, manage and maintain the conservation reserve.

(2) The committee shall consist of a representative of the forest or Wild Life Department, who shall be the Member Secretary of the Committee, one representative of each Village Panchayat in whose jurisdiction the reserve is located, three representatives of non governmental organisations working in the field of wild life conservation and one representative each from the Department of Agriculture and Animal Husbandry.

(3) The Committee shall regulate its own procedure including the quorum.

Declaration and Management of Community Reserve

36C. *(1)* The State Government may, where the community or an individual has volunteered to conserve wild life and its habitat, declare any private or community land not comprised within a National Park, sanctuary or a conservation reserve, as a community reserve, for protecting fauna, flora and traditional or cultural conservation values and practices.

(2) The provisions of sub section *(2)* of section 18, sub sections *(2)*, *(3)* and *(4)* of section 27, sections 30, 32 and clauses *(b)* and *(c)* of section 33 shall, as far as may be, apply in relation to a community reserve as they apply in relation to a sanctuary.

(3) After the issue of notification under sub section *(1)*, no change in the land use pattern shall be made within the community reserve, except in accordance with a resolution passed by the management, committee and approval of the same by the State Government.

Community Reserve Management Committee

36D. *(1)* The State Government shall constitute a Community Reserve management committee, which shall be the authority responsible for conserving, maintaining and managing the community reserve.

(2) The committee shall consist of five representatives nominated by the Village Panchayat or where such Panchayat does not exist by the members of the Gram Sabha and one representative of the State Forests or Wild Life Department under whose jurisdiction the community reserve is located.

(3) The committee shall be the competent authority to prepare and implement the management plan for the community reserve and to take steps to ensure the protection of wild life and its habitat in the reserve.

(4) The committee shall elect a Chairman who shall also be the Honorary Wild Life Warden on the community reserve,

(5) The committee shall regulate its own procedure including the quorum.".

Omission of Section 37

21. Section 37 of the principal Act shall be omitted.

Amendment of Section 38B

22. In section 38B of the principal Act, In sub section *(l)* after the words "every

member", the words "other than the Member Secretary" shall be inserted.

Amendment of Section 38H.

23. In section 38H of the principal Act, after sub section *(1)*, the following sub section shall be inserted, namely:—

"*(1A)* On and after the commencement of the Wild Life (Protection) Amendment Act, 2002 a zoo shall not be established without obtaining the prior approval of the Authority.".

Substitution of New Section for Section 38-1.

24. For section 38 1 of the principal Act, the following section shall be substituted, namely:—

Acquisition of Animals by a Zoo

"38 1. *(1)* Subject to the other provisions of this Act, no zoo shall acquire, sell or transfer any wild animal or captive animal specified in Schedules I and 11 except with the previous permission of the Authority.

(2) No zoo shall acquire, sell or transfer any wild or captive animal except from or to a recognised zoo.".

Amendment of Section 40

25. In section 40 of principal Act,—

(i) in sub section *(1)*, for the words "or any uncured trophy", the words "or animal article, trophy or uncured trophy" shall be substituted;

(ii) after sub section *(2)*, the following sub sections shall be inserted, namely:–

"*(2A)* No person other than a person having a' certificate of ownership, shall, after the commencement of the Wild Life (Protection) Amendment Act, 2002 acquire, receive, keep in his control, custody or possession any captive animal, animal article, trophy or uncured trophy specified in Schedule I or Part II of Schedule II, except by way of inheritance.

(2B) Every person inheriting any captive animal, animal article, trophy or uncured trophy under sub section *(2A)* shall, within ninety days of such inheritance make a declaration to the Chief Wild Life Warden or the authorised officer and the provisions of sections 41 and 42 shall apply as if the declaration had been made under sub section *(1)* of section 40:

Provided that nothing in sub sections *(2A)* and *(2B)* shall apply to the live elephant.".

(iii) in sub section *(4)*, for the words "any animal article", the words "any animal or animal article" shall be substituted.

Insertion of New Section 40A

26. After section 40 of the principal Act, the following section shall be inserted, namely:–

Immunity in Certain Cases

"40A. (1) Notwithstanding anything contained in sub sections *(2)* and *(4)* of section 40 of this Act, the Central Government may, by notification, require any person to declare to the Chief Wild Life Warden or the authorised officer, any captive animal, animal article, trophy or uncured trophy derived from animal, specified in Schedule I or Part 11 of Schedule 11 in his control, custody or possession, in respect of which no declaration had been made under sub section *(1)* or sub section *(4)* of section 40, in such form, in such manner and within such time as may be prescribed.

(2) Any action taken or purported to be taken for violation of section 40 of this Act at any time before the commencement

of the Wild Life (Protection) Amendment Act, 2002 shall not be proceeded with and all pending proceedings shall stand abated.

(3) Any captive animal, animal article, trophy or uncured trophy declared under sub section *(1)*, shall be dealt with in such manner and subject to such conditions as may be prescribed.".

Amendment of Section 42

27. In section 42 of the principal Act, the following proviso shall be added, namely:

"Provided that before issuing the certificate of ownership in respect of any captive animal, the Chief Wild Life Warden shall ensure that the applicant has adequate facilities for housing, maintenance and upkeep of the animal.".

Substitution of New Section for Section 43

28. For section 43 of the principal Act, the following section shall be substituted, namely .

Regulation of Transfer of Animal, etc.

"43. *(1)* No person having in his possession captive animal, animal article, trophy or uncured trophy in respect of which he has a certificate of ownership shall transfer by way of sale or offer for sale or by any other mode of consideration of commercial nature, such animal or article or trophy or uncured trophy.

(2) Where a person transfers or transports from the State in which he resides to another State or acquires by transfer from outside the State, any such animal, animal article, trophy or uncured trophy in respect of which he has a certificate of ownership, he shall, within thirty days of the transfer or transport, report the transfer or transport to the Chief Wild Life Warden or the authorised officer within whose jurisdiction the transfer or transport is effected.

(3) Nothing in this section shall apply–

(a) to tail feather of peacock and the animal article or trophies made therefrom;

(b) to transfer of captive animals between recognised zoos subject to the provisions of section 38-1, and transfer amongst zoos and public museums.".

Amendment of Section 50

29. In section 50 of the principal Act.

(i) in sub section *(3A)*, for the words "Wild Life Warden," the words "an of section 50. Assistant Conservator of Forests" shall be substituted;

(ii) in sub section *(4)* the words "under intimation to the Chief Wild Life Warden or the officer authorised by him in this regard" shall be inserted at the end;

(iii) for sub section *(6)*, the following sub section shall be substituted, namely:–

"*(6)* Where any meat, uncured trophy, specified plant or part or derivative thereof is seized under the provisions of this section, the Assistant Director of Wild Life Preservation or any other officer of a gazetted rank authorised by him in this behalf or the Chief Wild Life Warden or the authorised officer may arrange for the disposal of the same in such manner as may be prescribed.";

(iv) in sub section (8), for the words "Wild Life Warden", the words "an officer not below the rank of Assistant Conservator of Forests authorised by the State Government in this behalf shall be substituted.

Amendment of Section 51

30. In section 51 of the principal Act,–

(i) in sub section *(1)*, for the first and second provisos, the following provisos shall be substituted, namely:–

"Provided that where the offence committed is in relation to any animal specified in Schedule I or Part II of Schedule II or meat of any such animal or animal article, trophy or uncured trophy diri4ed from such animal or where the offence relates to hunting in a sanctuary or a National Park or altering the boundaries of a sanctuary or a National Park, such offence shall be punishable with imprisonment for a term which shall not be less than three years but may extend to seven years and also with fine which shall not be less than ten thousand rupees:

Provided further that in the case of a second or subsequent offence of the nature mentioned in this sub section, the term of the imprisonment shall not be less than three years but may extend to seven years and also with fine which shall not be less than twenty five thousand rupees.";

(ii) in sub section *(1A)*, for the words "one year ", the words "three years" and for the words "five thousand rupees", the words "ten thousand rupees" shall be substituted.

Insertion of New Section 51A

31. After section 51 of the principal Act, the following section shall be inserted, new section namely:—

Certain Condition to Apply while Granting Bail

2 of 1974

"51A. When any person accused of, the commission of any offence relating to Schedule I or Part II of Schedule If or offences relating to hunting inside the boundaries of National Park or wild life sanctuary or altering the boundaries of such parks and sanctuaries, is arrested under the provisions of the Act, then notwithstanding anything contained in the Code of Criminal Procedure, 1973 no such person who had been previously convicted of an offence under this Act shall, be released on bail unless—

(a) the Public Prosecutor has been given an opportunity of opposing the release on bail, and

(b) where the Public Prosecutor opposes the application, the Court is satisfied that there are reasonable grounds for believing that he is not guilty of such offence and that he is not likely to commit any offence while on bail.".

Substitution of New Section for Section 54

32. For section 54 of the principal Act, the following section shall be substituted, namely:–

Power of Compound Offences

"54, *(1)* The Central Government may, by notification, empower the Director of Wild Life Preservation or any other officer not below the rank of Assistant Director of Wild Life Preservation and in the case of a State Government in the similar manner, empower the Chief Wild Life Warden or any officer of a rank not below the rank of a Deputy Conservator of Forests, to accept from any person against whom a reasonable suspicion exists that he has committed an offence against this Act, payment of a sum of money by way of composition of the offence which such person is suspected to have committed.

(2) On payment of such sum of money to such officer, the suspected person, if in custody, shall be discharged and no further proceedings in respect of the offence shall be taken against such person.

(3) The officer compounding any offence may order the cancellation of any licence or permit granted under this Act to the offender, or if not empowered to do so, may approach an officer so empowered, for the cancellation of such licence or permit.

(4) The sum of money accepted or agreed to be accepted as composition under sub section (1) shall, in no case, exceed the sum of twenty five thousand rupees:

Provided that no offence, for which a minimum period of imprisonment has been prescribed in section 51, shall be compounded."

Amendment of Section 55

33. In section 55 of the principal Act,–

(i) after clause *(a)*, the following clause shall be inserted, namely:–"*(aa)* the Member Secretary, Central Zoo Authority in matters relating to violation of the provisions of Chapter IVA; or";

(ii) in clause *(b)*, after the words "State Government", the words "subject to such conditions as may be specified by that Government" shall be inserted;

(iii) after clause *(b)*, the following clause shall be inserted, namely: -

"*(bb)* the officer in charge of the zoo in respect of violation of provisions of section 38J; or".

INSERTION OF NEW CHAPTER VI A

34. After Chapter VI of the principal Act, the following Chapter shall be inserted, namely:–

FORFEITURE OF PROPERTY DERIVED FROM ILLEGAL HUNTING AND TRADE

Application

58A. The provisions of this Chapter shall apply only to the following persons, namely:–

(a) every person who has been convicted of an offence punishable under this Act with imprisonment for a term of three years or more;

(b) every associate of a person referred to in clause *(a)*;

(c) any holder (hereafter in this clause referred to as the present holder) of any property which was at any time previously held by a person referred to in clause *(a)* or clause *(b)* unless the present holder or, as the case may be, any one who held such property after such person and before the present holder, is or was a transferee in good faith for adequate consideration.

Definition

58B. In this Chapter, unless the context otherwise requires,–

(a) "Appellate Tribunal" means the Appellate Tribunal for Forfeited Property constituted under section 58N;

(b) "associate" in relation to a person whose property is liable to be forfeited under this Chapter, includes–

(*i*) any individual who had been or is managing the affairs or keeping the accounts of such person;

1 of 1956

(*ii*) any association of persons, body of individuals, partnership firm or private company within the meaning of the Companies Act 1956 of which such person had been or is a member, partner or director;

(*iii*) any individual who had been or is a member, partner or director of an association of persons, body of individuals, partnership firm or private company referred to in sub clause (*ii*) at any time when such person had been or is a member, partner or director of such association, body, partnership firm or private company;

(*iv*) any person, who had been or is managing the affairs, or keeping the accounts of any association of persons, body of individuals, partnership firm or private company referred to in sub clause (*iii*);

(*v*) the trustee of any trust, where–

(*1*) the trust has been created by such person; or

(*2*) the value of the assets contributed by such person (including the value of the assets, if any, contributed by him earlier) to the trust amounts on the date on which contribution is made, to not less than twenty per cent of the value of the assets of the trust on that date;

(*vi*) where the competent authority, for reasons to be recorded in writing, considers that any properties of such person are held on his behalf by any other person, such other person;

(*c*) "competent authority" means an officer authorised under section 58D;

(*d*) "concealment" means the concealment or disguise of the nature, source disposition, movement or ownership of property and includes the movement or conversion of such property by electronic transmission or by any other means;

(e) "freezing" means temporarily prohibiting the transfer, conversion, disposition or movement of property by an, order issued under section 58F;

(f) "identifying' includes establishment of proof that the property was derived from, or used in, the illegal hunting and trade of wild life and its products;

(g) "illegally acquired property" in relation to any person to whom this Chapter applies, means–

(i) any property acquired by such person, wholly or partly out of or by means of any income earnings or assets derived or obtained from or attributable to illegal hunting and trade of wild life and its products and derivatives;

(ii) any property acquired by such person, for a consideration or by any means, wholly or partly traceable to any property referred to in sub-clause *(i)* or the income or earning from such property, and includes–

(A) any property held by such person which would have been, in relation to any previous holder thereof, illegally acquired property under this clause if such previous holder had not ceased to hold it, unless such person or any other person who held the property at any time after such previous holder or, where there are two or more such previous holders, the last of such previous holders is or was a transferee in good faith for adequate consideration;

(B) any property acquired by such person, for a consideration, or by any means, wholly or partly traceable to any property failing under item *(A)*, or the income or earnings therefrom;

(h) "property' means property and assets of every description, whether corporeal or incorporeal, movable or immovable, tangible or intangible and deeds and instruments evidencing title to, or interest in, such property or assets, derived from the illegal hunting and trade of wild life and its products;

(i) "relative" means–

(1) spouse of the person;

(2) brother or sister of the person;

(3) brother or sister of the spouse of the person;

(4) any lineal ascendant or descendant of the person,

(5) any lineal ascendant or descendant of the spouse of the person;

(6) spouse of a person referred to in sub–clause *(2)*, sub clause *(3)*, sub clause *(4)* or sub clause *(5)*;

(7) any lineal descendant of a person referred to in sub clause *(2)* or sub clause *(3)*;

(j) "tracing" means determining the nature, source, disposition, movement, title or ownership of property;

(k) "trust" includes any other legal obligation.

Prohibition of Holding Illegally Acquired Property

58C. *(1)* From the date of commencement of this Chapter, it shall not be lawful for any person to whom this Chapter applies to hold any illegally acquired property either by himself or through any other person on his behalf.

(2) Where any person holds such property in contravention of the provisions of sub section *(1)*, such property shall be liable to be forfeited to the State Government concerned in accordance with the provisions of this Chapter:

Provided that no property shall be forfeited under this Chapter if such property was acquired by a person to whom this Act applies before a period of six years from the date on which he was charged for an offence relating to illegal hunting and trade of wild life and its products.

Competent Authority

58D. The State Government may, by order published in the Official Gazette, authorise any officer not below the rank of Chief Conservator of Forests to perform the functions of the competent authority under this Chapter in respect of such persons or classes of persons as the State Government may, direct.

Identifying Illegally Acquired Property

58E. *(1)* An officer not below the rank of Deputy Inspector General of Police duly authorised by the Central Government or as the case may be, the State Government, shall, on receipt of a complaint from the competent authority about any person having illegally acquired property, proceed to take all steps necessary for tracing and identifying any property illegally acquired by such person.

(2) The steps referred to in sub section *(1)* may include any inquiry, investigation or survey in respect of any person, place, property, assets, documents, books of account in any bank or financial institution or any other relevant step as may be necessary.

(3) Any inquiry, investigation or survey referred to in sub section *(2)* shall be carried out by an officer mentioned in sub section *(1)* in accordance with which directions or guidelines as the competent authority may make or issue in this behalf.

Seizure or Freezing of Illegally Acquired Property

58F. *(1)* Where any officer conducting an inquiry or investigation under section 58E has mason to believe that any property in relation to which such inquiry or investigation is being conducted is an illegally acquired property and such property is likely to be concealed, transferred or dealt with in any manner which may result in frustrating any proceeding relating to forfeiture of such property under this Chapter, he may make an order for seizing such property arid where it is not practicable to seize such property, he may make an order that such property shall not be transferred or otherwise dealt with, except with the prior permission of the officer making such order, or of the competent authority arid a copy of such order shall be served on the person concerned:

Provided that a copy of such an order shall be sent to the competent authority within forty eight hours of its being made.

(2) Any order made under sub section *(1)* shall have no effect unless the said order is confirmed by an order of the

competent authority within a period of thirty days of its being made.

Explanation. For the purposes of this section, "transfer of property" means any disposition, conveyance, assignment, settlement, delivery, payment or other alienation of property and, without limiting the generality of the foregoing, includes–

(a) the creation of a trust in property;

(b) the grant or creation of any lease, mortgage, charge, easement, licence, power, partnership or interest in property;

(c) the exercise of a power of appointment, of property vested in any person not the owner of the property, to determine its disposition in favour of any person other than the donee of the power; and

(d) any transaction entered into by any person with intent thereby to diminish directly or indirectly the value of his own property and to increase the value of the property of any other person.

Management of Properties Seized or Forfeited under this Chapter

58G. *(1)* The State Government may, by order published in the Official Gazette, appoint as many of its officers (not below the rank of Conservator of Forests) as it thinks fit, to perform the functions of an Administrator.

(2) The Administrator appointed under sub section *(1)* shall receive and manage the property in relation to which an order has been made under sub section *(1)* of section 58F or under section 58 1 in such manner and subject to such conditions as may be prescribed.

(3) The Administrator shall also take such measures as the State Government may direct, to dispose of the property which is forfeited to the State Government.

Notice of Forfeiture of Property

58H. *(1)* If having regard to the value of the properties held by any person to whom this Chapter applies, either by himself or through any, other person on his behalf, his known sources of income, earnings or assets, and any other information or material available to it as a result of a report from any officer making an investigation under section 58E or otherwise, the competent authority for reasons to be recorded in writing believes that all or any of such properties are illegally acquired properties, it may serve a notice upon such person (hereinafter referred to as the person affected) calling upon him within a period of thirty days specified in the notice to show cause why all or any of such properties, as the case may be, should not be declared to be illegally acquired properties and forfeited to the State Government under this, Chapter and in support of his case indicate the sources of his income, earnings or assets, out of which or by means of which he has acquired such property, the evidence on which he relies and other relevant information and particulars.

(2) Where a notice under sub section *(1)* to any person specifies any property as being held on behalf of such person by any other person, a copy of the notice shall also be served upon such other person.

Forfeiture of Property in Certain Cases

58 I *(1)* The competent authority may, after considering the explanation, if any, to the show cause notice Issued under section 58H, and the materials available before it and after giving to the person affected and in a case where the person affected holds any property specified in the notice through any other person, to such other person, also a reasonable opportunity of being heard, by order, record a finding whether al I or any of the properties in question are illegally acquired properties:

Provided that if the person affected (and in a case where the person affected holds any property specified in the notice through any other person, such other person also), does not appear before the competent authority or represent his case before it within a period of thirty days specified in the show cause notice, the competent authority may proceed to record a finding under this sub section *ex parte* on the basis of evidence available before it.

(2) Where the competent authority is satisfied that some of the properties referred to in the show cause notice are illegally acquired properties but is not able to identify specifically such properties, then, it shall be lawful for the competent authority to specify the properties which, to the best of its judgment, are illegally acquired properties and record a finding accordingly under sub section *(1)* within a period of ninety days.

(3) Where the competent authority records a finding under this section to the effect that any property is illegally acquired property, it shall declare that such property shall, subject to the provisions of this Chapter stand forfeited to the State Government free from all encumbrances.

(4) In case the person affected establishes that the property specified in the notice issued under section 58H is not an illegally acquired property and therefore not liable to be forfeited under the Act, the said notice shall be withdrawn and the property shall be released forthwith.

(5) Where any shares in a company stand forfeited to the State Government under this, Chapter, the company shall, notwithstanding anything contained in the Companies Act, 1956 or the article of association of the company, forthwith register the State Government as the transferee of such shares.

Burden of Proof

58J. In any proceedings under this Chapter, the burden of proving that any property specified in the notice served under section 59H is not illegally acquired property shall be on the person affected.

File in Lieu of Forfeiture

58K. *(1)* Where the competent authority makes, a declaration that any property stands forfeited to the State Government under section 58 1 and it is a case where the source of only a part of the illegally acquired property has not been proved to the satisfaction of the competent authority, it shall make an order giving option to the person affected to pay, in lieu of forfeiture, a fine equal to the market value of such part.

(2) Before making an order imposing a fine under sub section *(1)*, the person affected shall be given a reasonable opportunity of being heard.

(3) Where the person affected pays the fine due under sub section *(1)*, within such time as may be allowed in that behalf, the competent authority may, by order revoke the declaration of forfeiture under section 58 1 and thereupon such property shall stand released.

Procedure in relation to certain Trust Properties

58L. In the case of any person referred to in sub clause *(vi)* of clause *(b)* of section 58B, if the competent authority, on the basis of the information and materials available to it, for reasons to be recorded in writing believes that any property held in trust is illegally acquired property, it may serve a notice upon the author of the trust, or as the case may be, the contributor of the assets out of or by means of which such property was acquired by the trust and the trustees, calling upon them within a period of thirty days specified in the notice, to explain the source of money or other asset out of or by means of which such property was acquired or, as the case may be, the source of money or other assets which were contributed to the trust for acquiring such property and thereupon such notice shall be deemed to be a notice served under section 58H and all the other provisions of this Chapter shall apply accordingly.

Explanation.–For the purposes of this section "illegally acquired property" in relation to any property held in trust, includes—

(i) any property which if it had continued to be held by the author of the trust or the contributor of such property to the trust would have been illegally acquired property in relation to such author or contributor;

(ii) any property acquired by the trust out of any contributions made by any person which would have been

illegally acquired property in relation to such person had such person acquired such property out of su6 contributions.

Certain Transfers to be Null and Void

58M. Where after the making of an order under sub section *(1)* of section 58F or the issue of a notice under section 58k or under section 58L, any property referred to in the said order or notice is transferred by any mode whatsoever, such transfer shall, for the purposes of the proceedings under this Chapter, be ignored and if such property is subsequently forfeited to the State Government under section 584, then, the transfer of such property shall be deemed to be null and void.

Constitution of Appellate Tribunal

58N. *(1)* The State Government may, by notification in the Official Gazette, constitute an Appellate Tribunal to be called the Appellate Tribunal for Forfeited Property consisting of a Chairman, and such number of other members (being officers of the State Government not below the rank of a Principal Secretary to the Government), as the State Government thinks fit, to be appointed by that Government for hearing appeals against the orders made under section 58F, section 58 I, sub section *(i)* of section 58K or section 58L.

(2) The Chairman of the Appellate Tribunal shall be a person who is or has been or is qualified to be a Judge of a High Court.

(3) The terms and conditions of service of the Chairman and other members shall be such as may be prescribed.

Appeals

58 O. *(1)* Any person aggrieved by an order of the competent authority made under section 58F, section 58 1, sub section *(1)* of section 58K or section 58L may, within forty five days from the date on which the order is served on him prefer an appeal to the Appellate Tribunal:

Provided that the Appellate Tribunal may entertain an appeal after the said period of forty five days, but not after sixty days, from the date aforesaid if it is satisfied that the appellant was prevented by sufficient cause from filing the appeal in time.

(2) On receipt of an appeal under sub section *(1)*, the Appellate Tribunal may, after giving an opportunity of being heard to the appellant, if he so desires, and after making such further inquiry as it deems fit, confirm, modify or set aside the order appealed against.

(3) The Appellate Tribunal may regulate its own procedure.

(4) On application to the Appellate Tribunal and on payment of the prescribed fee, the Appellate Tribunal may allow a party to any appeal or any person authorised in this behalf by such party to inspect at any time during office hours, any relevant records and registers of the Appellate Tribunal and obtain a certified copy or any part thereof.

Notice or Order not to be invalid for Error in Description

58P. No notice issued or served, no declaration made, and no order passed under this Chapter shall be deemed to be invalid by reason of any error in the description of the property or person mentioned therein if such property or person is identifiable from the description so mentioned.

Bar of Jurisdiction

58Q. No order passed or declaration made under this Chapter shall be appealable except as provided therein and no civil court shall have jurisdiction in respect of any matter which the Appellate Tribunal or any competent authority is empowered by or under this Chapter to determine, and no injunction shall be granted by any court or other authority in respect of any action taken or to be taken in pursuance of any power conferred by or under this Chapter.

Competent Authority and Appellate Tribunal to have Powers of Civil Court

58R. The competent authority and the Appellate Tribunal shall have all the powers of a civil court while trying a suit under the Code of Civil Procedure, 1908 in respect of the following matters, namely:–

(a) summoning and enforcing the attendance of any person and examining him on oath;

(b) requiring the discovery and production of documents;

(c) receiving evidence on affidavits;

(d) requisitioning any public record or copy thereof from any court or office;

(e) issuing commissions for examination of witnesses or documents;

(f) any other matter which may be prescribed

Information to Competent Authority

58S. *(1)* Notwithstanding anything contained in any other law for the time being in force, the competent authority shall have power to require any officer or authority of the Central Government or a State Government or a local authority to furnish information in relation to such persons, on points or matters as in the opinion of the competent authority will be useful for, or relevant to, the purposes of this Chapter.

(2) Every officer referred to in section 58T may furnish *suo motu* any information available with him to the competent authority if in the opinion of the officer such information will be useful to the competent authority for the purposes of this Chapter.

Certain Officers to Assist Administrator, Competent Authority and Appellate Tribunal

58T. For the purposes of any proceedings under this Chapter, the following officers shall render such assistance as may be necessary to the Administrator appointed under section 58G, competent authority and the Appellate Tribunal, namely:–

(a) officers of Police;

(b) officers of the State Forest Departments;

(c) officers of the Central Economic Intelligence Bureau;

(d) officers of the Directorate of Revenue Intelligence;

(e) such other officers as specified by the State Government in this behalf by notification in the Official Gazette.

Power to Take Possession

58U. *(1)* Where any property has been declared to be forfeited to the State Power to take Government under this Chapter, or where the person affected has failed to pay the fine due under sub section *(1)* of section 58K within the time allowed therefor under subsection *(3)* of that section, the competent authority may order the person affected as well as any other person who may be in possession of the property to surrender or deliver possession thereof to the Administrator appointed under section 58G or to any person duly authorised by him in this behalf within thirty days of the service of the order.

(2) If any person refuses or fails to comply with an order made under sub section *(1)*, the Administrator may take possession of the property and may for that purpose use such force as may be necessary.

(3) Notwithstanding anything contained in sub section *(2)*, the Administrator may, for the purpose of taking possession of any property referred to in sub section *(1)* requisition the services of any police officer to assist him and it shall be the duty of such officer to comply with such requisition.

Rectification of Mistakes

58V. With a view to rectifying any mistake apparent from record, the competent authority or the Appellate Tribunal, as the case maybe, may amend any order made by it within a period of one year from the date of the order:

Provided that if any such amendment is likely to affect any person prejudicially and the mistake is not of a clerical nature, it shall not be made without giving to such person a reasonable opportunity of being heard.

Findings under other Laws not Conclusive for Proceedings under this Chapter

58W. No finding of any officer or, authority under any other law shall be conclusive for the purposes of any proceedings under this Chapter.

Service of Notices and Orders

58X. Any notice or order issued or made under this Chapter shall be served,–

(a) by tendering the notice or order or sending it by registered post to the person for whom it is intended or to his agent;

(b) if the notice or order cannot be served in the manner provided in clause *(a)*, then, by affixing it on a conspicuous place in the property in relation to which the notice or order is issued or made or on some conspicuous part of the premises in which the person for whom it is intended is known to have last resided or carried on business or personally worked for gain.

Punishment for Acquiring Property in relation to which Proceedings have been taken under this Chapter

58Y. Any person who knowingly acquires, by any mode whatsoever, any property in relation to which proceedings are pending under this Chapter shall be punishable with imprisonment for a term Which may extend to five years and with fine which may extend to fifty thousand rupees.'.

Amendment of Section 60A

35. In section 60A of the principal Act,–

(i) in sub section *(1)*, for the words "twenty per cent of such fine", the words "fifty per cent. of such fine" shall be substituted;

(ii) in sub section *(2)*, for the words "twenty per cent of such money", the words "fifty per cent of such money" shall be substituted.

Insertion of New Section 60B

36. After section 60A of the principal Act, the following section shall be inserted, namely-:–

Reward by State Government

"60B.The State Government may empower the Chief Wild Life Warden to order payment of reward not exceeding ten thousand rupees to be paid to a person who renders assistance in the detection of the offence or the apprehension of the offender, from such fund and in such manner as may, be prescribed.".

Amendment of Section 63

37. In section 63 of the principal Act, in sub section *(1)*, the following clause shall be inserted, namely:–

"*(ai)* the term of office of members other than those who are members *ex officio*; the manner of filling vacancies, the procedure to be followed by the National Board under sub section *(2)* and allowances of those members under sub section, *(3)* of section 5A;".

Amendment of Section 64

38. In section 64 of the principal Act, in sub section (2),–

(i) for clauses *(a)* and *(b)*, the following clauses shall be substituted, namely:–

"*(a)* the term of office of members other than those who are members, *ex officio*, the manner of filling vacancies and the procedure to be followed by the Board under sub section *(2)* of section 6;

(b) allowances referred to in sub section *(3)* of section 6;";

(ii) after clause *(d)*, the following clause shall be inserted, namely:–

"*(dd)* the conditions subject to which the officers will be authorised to file cases in the court;";

"*(iii)* after clause *(g)*, the following clauses shall be inserted, namely:–

"*(ga)* the manner and conditions subject to which the Administrator shall receive and manage the property under sub section *(2)* of section 58G;

(gb) the terms and conditions of service of the Chairman and other members under sub-section *(3)* of section 58N;

(gc) the fund from which and the manner in which payment of reward under section 60B shall be made;".

Bibliography

Ausubel, F.M.: *Current Protocols in Molecular Biology,* New York, John Wiley and Sons, 1989.

Bernauer, Thomas: *Genes, Trade, and Regulation: The Seeds of Conflict in Food Biotechnology,* Princeton, Princeton University. 2003.

Brandwein, P.F.: *Sourcebook for the Biological Sciences,* San Diego, Harcourt Brace JOvanovich, 1986.

Broach, J.R.: *The Molecular Biology of the Yeast,* Cold Spring Harbor, Cold Spring Harbor Laboratory, 1981.

Chrispeels, Maarten : *Plants, Genes and Crop Biotechnology,* Sudbury MA, Jones and Barlett Publishers, 2003.

Costanza, R.: *Ecological Economics,* Columbia Uni. Press., New York, 1991.

Cox, Earl D. : *Beyond Humanity: Cyber Evolution and Future Minds,* Roackland, Charles River Media, 1996.

Daubenmire, F.: *Plants and Environment,* New York, Wiley, 1947.

David Sadava: *Plants, Genes and Crop Biotechnology,* Sudbury MA, Jones and Barlett Publishers, 2003.

Dodds, John H.: *Plant Genetic Engineering,* New York, Cambridge University Press, 1985.

Dusenbery, David B.: *Life at a Small Scale: the Behavior of Microbes,* New York, Scientific American Library, 1996.

E. Dereniak and D. Crowe, *Optical Radiation Detectord,* John Wiley and Sons, 1984.

Fransman M, Junne G, Roobeek A: *The Biotechnology Revolution?,* Oxford, Blackwell, 1995.

Jameson, J. L.: *Principles of Molecular Medicine,* Totowa, Humana Press, 1998.

Goodman, D.C.: *From Farming to Biotechnology: A Theory of Agro-industrial Development*, Oxford, Blackwell, 1987.

Lodish, H.: *Molecular Cell Biology*, New York, Freeman, 2000.

Maarten, C.: *Plants, Genes and Crop Biotechnology*, Sudbury MA, Jones and Barlett Publishers, 2003.

Murray, David: *Seeds of Concern: The Genetic Manipulation of Plants*, Sydney, University of New South Wales, 2003.

Postman, Neil: *Technopoly: The Surrender of Culture to Technology*, New York, Vintage Books, 1992.

Raff, K. Roberts, and J. Watson, *Molecular Biology of the Cell*, New York: Garland Publishing, 1989.

Sheldon, J.K.: *Practical Environmental Bioremediation*, Lewis Publishers, Boca Raton, 1992.

Shetty, Kalidas: *Food Biotechnology*, New York, Dekker/CRC Press, 2005.

Slater, R.J.: *Experiments in Molecular Biology*, Clifton, Humana Press, 1986.

Teich, M.: *Fundamentals of Photonics*, New York: Wiley, 1991.

Walden, Richard: *Genetic Transformation in Plants*, England, Open University Press, 1988.

Wood, E.J. Ferguson: *Marine Microbial Ecology*, London, Chapman and Hall, 1965.

INDEX

A

Accounts, 46, 47, 70, 161, 163, 205, 206, 207, 208, 213, 219, 220, 221, 245.

Animal Trade, 58.

Animals, 1, 2, 6, 8, 9, 11, 12, 13, 14, 20, 22, 23, 28, 29, 32, 33, 39, 45, 48, 49, 51, 54, 58, 59, 60, 72, 79, 81, 85, 87, 88, 89, 90, 91, 92, 93, 94, 97, 100, 101, 102, 106, 112, 113, 115, 122, 190, 192, 211, 222, 224, 231, 238, 241.

Application, 31, 32, 38, 40, 47, 48, 53, 54, 55, 71, 72, 73, 78, 79, 80, 81, 86, 87, 90, 94, 99, 108, 109, 111, 190, 191, 193, 194, 200, 201, 212, 215, 219, 242, 244, 254.

Appointment, 25, 44, 70, 77, 112, 113, 114, 166, 179, 197, 232, 249.

Authority, 39, 43, 44, 45, 46, 47, 48, 56, 69, 70, 86, 87, 90, 91, 92, 93, 94, 100, 112, 117, 118, 119, 128, 139, 141, 147, 168, 170, 171, 172, 173, 175, 176, 180, 182, 184, 187, 188, 191, 192, 193, 194, 195, 196, 197, 198, 199, 200, 201, 202, 203, 205, 206, 207, 209, 210, 211, 212, 213, 214, 215, 216, 217, 219, 220, 221, 227, 237, 238, 243, 245, 247, 248, 249, 250, 251, 252, 253, 254, 255, 256, 257.

C

Captive Animals, 23, 45, 51, 54, 60, 72, 97, 241.

Committee, 18, 77, 118, 156, 157, 158, 165, 166, 167, 168, 169, 170, 178, 179, 180, 181, 182, 197, 198, 200, 201, 211, 213, 228, 234, 236, 237, 238.

Conservation, 2, 28, 48, 76, 87, 112, 113, 115, 117, 118, 120, 138, 155, 161, 165, 166, 176, 178, 179, 188, 189, 195, 196, 199, 203, 204, 206, 207, 208, 209, 210, 211, 212, 213, 221, 224, 228, 234, 236, 237.

Conservator, 66, 105, 166, 169, 170, 172, 173, 174, 175, 181, 182, 184, 187, 234, 241, 243, 247, 249.

Consideration, 76, 105, 106, 159, 167, 180, 192, 240, 244, 246.

D

Department, 3, 17, 26, 27, 117, 119, 128, 156, 166, 175, 187, 189, 226, 227, 229, 230, 236, 237.
Detection, 61, 63, 69, 115, 116, 258.
Diversity, 156, 187, 188, 189, 190, 191, 192, 195, 196, 198, 203, 208, 209, 211.

E

Endangerment, 1, 2.
Environment, 3, 33, 36, 41, 84, 88, 89, 165, 166, 169, 171, 176, 179, 181, 187, 195.
Exports, 99, 100, 101, 102, 105.

F

Finance, 205, 207, 227.
Forest Conservation Rules, 178.
Forest Department, 17, 26, 27, 156, 175, 187, 229.
Forests, 1, 13, 15, 26, 66, 84, 105, 113, 117, 118, 121, 122, 130, 131, 132, 134, 135, 136, 137, 138, 152, 155, 156, 158, 161, 165, 166, 169, 170, 172, 173, 174, 175, 176, 179, 181, 182, 184, 187, 195, 218, 226, 227, 228, 229, 234, 237, 241, 243, 247, 249.

G

Government, 2, 16, 19, 20, 21, 22, 25, 26, 27, 28, 29, 30, 31, 32, 33, 35, 36, 37, 39, 40, 41, 42, 43, 44, 46, 47, 48, 49, 50, 55, 56, 57, 59, 60, 63, 65, 66, 67, 69, 70, 71, 73, 74, 75, 77, 78, 79, 80, 81, 83, 84, 85, 86, 99, 100, 106, 107, 112, 114, 116, 117, 118, 119, 121, 122, 123, 124, 125, 127, 128, 129, 130, 131, 132, 133, 134, 136, 137, 138, 139, 140, 141, 142, 143, 144, 145, 146, 147, 148, 149, 150, 151, 152, 153, 154, 156, 157, 158, 159, 160, 161, 162, 163, 165, 166, 167, 168, 169, 170, 171, 172, 175, 176, 177, 178, 179, 180, 181, 182, 188, 190, 191, 193, 195, 196, 198, 199, 200, 202, 203, 204, 205, 206, 207, 208, 209, 210, 211, 212, 213, 214, 216, 217, 218, 219, 220, 221, 222, 223, 224, 226, 227, 228, 229, 230, 231, 232, 233, 234, 235, 236, 237, 239, 241, 243, 244, 247, 248, 249, 250, 251, 253, 255, 256, 258.
Guidelines, 92, 112, 188, 193, 199, 202, 204, 248.

I

Imports, 99, 100.
Indian Forest Act, 20, 21, 121, 156, 158, 163, 223, 224.

J

Justice, 189, 214.

L

Law, 28, 46, 59, 61, 63, 67, 99, 117, 121, 145, 151, 153, 160, 167, 179, 180,

189, 192, 194, 205, 207, 210, 212, 218, 227, 234, 255, 257.
Legislative, 189, 229.

M

Management, 29, 39, 42, 45, 94, 118, 132, 135, 137, 152, 156, 157, 160, 168, 173, 181, 184, 192, 193, 199, 207, 210, 211, 212, 213, 221, 228, 230, 233, 234, 235, 236, 237, 249.

N

National Biodiversity Authority, 187, 188, 191, 192, 193, 194, 195, 196, 197, 198, 199, 200, 201, 202, 203, 205, 206, 207, 209, 210, 211, 212, 213, 214, 215, 216, 217, 219, 220, 221.
National Park, 5, 11, 17, 21, 33, 40, 41, 42, 43, 49, 62, 64, 65, 73, 74, 113, 172, 181, 183, 224, 231, 234, 235, 237, 242.

P

Policy, 28, 99, 100, 193, 213, 214.
Powers, 25, 26, 34, 36, 43, 64, 66, 69, 74, 78, 81, 83, 84, 89, 105, 107, 112, 115, 116, 124, 141, 149, 150, 162, 165, 178, 188, 196, 199, 203, 215, 216, 219, 220, 228, 255.
Property, 20, 28, 29, 33, 37, 49, 50, 63, 66, 81, 113, 115, 122, 133, 136, 141, 142, 144, 145, 146, 147, 148, 149, 152, 153, 154, 193, 195, 200, 202, 203, 244, 245, 246, 247, 248, 249, 250, 251, 252, 253, 254, 256, 257, 259.
Protection, 12, 19, 27, 28, 30, 31, 32, 47, 58, 67, 68, 79, 90, 93, 94, 100, 101, 102, 103, 106, 113, 121, 125, 126, 128, 130, 132, 133, 134, 135, 149, 151, 152, 153, 155, 176, 178, 179, 180, 183, 184, 194, 212, 236, 249, 256, 259.

R

Regulation, 51, 72, 99, 100, 101, 136, 137, 140, 192, 218, 220, 240.

S

Secretary, 26, 27, 43, 44, 46, 68, 69, 70, 75, 76, 77, 84, 85, 94, 166, 169, 175, 179, 181, 184, 218, 226, 227, 228, 229, 230, 234, 236, 238, 243, 253.

W

Wild Animals, 28, 29, 33, 39, 45, 49, 100, 101, 106, 112, 113, 115, 122, 222.
Wildlife Protection Act, 19.
Wildlife Specified Plants, 83.

Z

Zoo Authority, 43, 46 70, 86, 87, 90, 91, 92, 93, 94, 227, 243.

□□□